Consciousness, Attention, and Conscious Attention

Consciousness, Attention, and Conscious Attention

Carlos Montemayor and Harry Haroutioun Haladjian

The MIT Press
Cambridge, Massachusetts
London, England

© 2015 Massachusetts Institute of Technology

All rights reserved. No part of this book may be reproduced in any form by any electronic or mechanical means (including photocopying, recording, or information storage and retrieval) without permission in writing from the publisher.

MIT Press books may be purchased at special quantity discounts for business or sales promotional use. For information, please e-mail special_sales@mitpress.mit.edu

This book was set in Stone by the MIT Press. Printed and bound in the United States of America.

Library of Congress Cataloging-in-Publication Data
Montemayor, Carlos.
Consciousness, attention, and conscious attention / Carlos Montemayor and Harry Haroutioun Haladjian.
　pages　cm
Includes bibliographical references and index.
ISBN 978-0-262-02897-4 (hardcover : alk. paper) 1. Consciousness. 2. Attention. I. Haladjian, Harry Haroutioun, 1973– II. Title.
BF311.M595 2015
153.7'33—dc23

2014034378

10 9 8 7 6 5 4 3 2 1

Contents

Preface   vii
Acknowledgments   xiii

1   Introduction   1
2   Forms of Attention   25
3   Forms of Consciousness   85
4   Conscious Attention   141
5   Consciousness–Attention Dissociation and the Evolution of Conscious Attention   177

Notes   217
Glossary   229
References   235
Index   275

Preface

In any type of mental activity, a person selects and analyzes various forms of information. Out of many possible color experiences, you see a specific set of highly unique color shades at any given time; out of many possible beliefs and desires, only a concrete subset of beliefs and desires leads you to act at any given moment. This process of selection, which is highly contextual, is responsible for determining the contents that your mind presents to you, in different varieties of mental states, from perception and dreams to inferential reasoning and beliefs about your past and future. In some of these activities, rules and routines direct your attention to certain contents, such as when you look at a work of art from different angles and distances in order to fully appreciate its aesthetic details, or when you follow step by step someone's train of thought in order to understand their intricate argument. In other cases, salient features of aspects of the world grab your attention, accompanied by memories or emotions you may have at that time, such as when you see a funnel-web spider near your foot and intense fear freezes you into inaction, or when a bright red light in the darkness captures your attention.

The more varied are the modes and perspectives you can take in selecting information, the richer the contents and implications of these selective processes will be. For example, to use the language of phenomenology: if you are capable of varying the procedures, perspectives, and contents you engage, you can take a more transcendental perspective and "bracket" your responses to and interpretations of your natural reactions to these contents, in order to focus your attention on how they *appear* to you. These contents include the manner in which your own perspective as a self appears to you. Such a process of focusing on how contents appear in your conscious awareness is basic in any exercise in which the focus of your attention is on how your experience is presented or given to you, such as in meditation.

The selection of highly specific and integrated contents, which may be voluntary or involuntary, is typical of conscious awareness. Since in all these mental activities of selection, focusing, and orientation you seem to be attending to different contents, it is tempting to stipulate that consciousness is the same as attention; attention seems to specify what conscious awareness is, or alternatively, all cases of attention are cases of conscious attention. This book challenges that intuitive idea. The empirical evidence and a careful analysis of philosophical and theoretical considerations reveal that in many circumstances attention is dissociated from consciousness.

We do not deny, however, that the changes in attention just mentioned are characteristic of conscious awareness. In fact, we propose that the focusing of attention on the highly specific contents of awareness—and its richly varied results—depend on the overlap between consciousness and attention. This overlap, *conscious attention*, is involved in the intricate ways in which one guides and connects the contents of experience associated with phenomenology. By contrast, unconscious attention is fundamental in selection processes that structure basic features of perceptual scenes and representational frameworks at the earliest stages of information processing. While unconscious attention implicitly informs conscious attention, it occurs without our being aware of such processes. Metaphorically speaking, if attention is a collection of context-dependent preprogrammed instructions to focus and refocus the mechanism responsible for the selection of information, or a collection of principles for the selection of information, consciousness seems to be independent of any set of such instructions or principles. More specifically, while attention operates by selecting and analyzing, consciousness operates by integrating and unifying. This book will explain such differences between consciousness and attention, while also elaborating on the theoretical implications of their dissociation.

An important and related idea we develop is that attention is deeply associated with epistemic processes (i.e., cognitive processes that lead to knowledge of facts), while consciousness is deeply associated with more social processes such as empathy. Conscious awareness helps us appreciate the experiences of others. We explain why this way of understanding attention and consciousness is implicitly assumed in many extant views on these topics. Our arguments elucidate the multiple ways in which epistemic processes interact with consciousness, producing the great variety of conscious awareness that human beings enjoy.

Unlike most contemporary analyses of consciousness, which focus mainly on perception, we also apply our theoretical approach to memory and expertise. These applications are particularly insightful because they illustrate the validity and generality of the dissociation between consciousness and attention. With respect to memory, we show how this dissociation helps explain that memories are important—not only because they afford reliable information about the past, but also because of what they evoke in us. With respect to expertise, we explain how this dissociation operates in situations in which one performs a very challenging task that one is well trained to do: the experience of performing such tasks, in spite of the challenging informational demands, is one of pleasant selflessness without the effortful attention that seemed necessary when first learning to perform such tasks (e.g., playing a musical instrument).

The empirical evidence and the theoretical views on attention reveal, moreover, that its dissociation from consciousness can partly be accounted for due to the great variety of attentional processes that occur at different levels of information processing, from basic feature detection to face recognition and expertise. These different forms of attention, we argue, are likely to have evolved at different times. Dissociation is a natural way of understanding the evolutionary gradation of attention, from strictly selective and automatic processes to complex crossmodal, voluntary, and semantically driven effortful attention. Furthermore, just as there is a gradation of attention, so there is a gradation of consciousness, from basic forms of awareness, like experiencing pain and hunger, to reflective forms of self-awareness, such as the desire, after receiving bad news, to believe that the information just learned is false.

There is a rich variety of ways in which we attend to and become consciously aware of information, and such variety is best understood in terms of the dissociation we propose. The varieties of attention have been studied extensively by cognitive psychologists and have led some of them to conclude that attention cannot be a single phenomenon. Consciousness, too, has often been dismissed by psychologists and philosophers as a broad phenomenon that may not correspond to a single kind of process. While we agree that one should not use the terms 'consciousness' or 'attention' without qualification, we disagree with such skepticism. We believe that attention can be viewed as a uniform, yet gradated phenomenon, characterized by its epistemic and analytic functions. Although the varieties of consciousness have been studied mostly by philosophers, they are now receiving the intense empirical study they deserve. We believe that, analogously to attention, different theories and forms of conscious awareness characterize a single phenomenon, which can be generally described as an empathetic and possibly integrative process.

There is also purpose in this rich variety. The most basic forms of early attention are structural and mechanically responsive; they have a long evolutionary history. Yet, their interactions with voluntary and semantically oriented forms of attention reveal that all kinds of attention help integrate information for knowledge and action. In the case of conscious awareness, the main result is the vivid engagement of a subject with respect to an experience. In both cases, dissociation facilitates variation. We do not make the claim that there is intrinsic purposeful action in nature; however, we believe that conscious attention evolved after the basic forms of attention, and that its evolution increased the access we have to the richest kinds of cognitive contents.

Finally, parsimony is important, and it has in previous accounts been used to challenge the dissociation between consciousness and attention. Would it not be simpler and more economical to say that consciousness just is attention? Throughout the book, we not only explain why dissociation is theoretically and empirically adequate, but show that it actually is parsimonious: not all forms of consciousness are linked to the varied forms of attention, making cognitive processing much more efficient than if consciousness and attention were a single phenomenon. In fact, we think that the only way to make real progress in the study of consciousness is by fully comprehending how deeply important its dissociation from attention is. You will find detailed arguments that demonstrate this point in the chapters that follow. But to dissuade those who associate variety with lack of parsimony, we believe it suffices to recall the basic principle of the theory of evolution: that complexity and diversity increase as simple principles remain in place, allowing for a perfect balance between variation and parsimony.

The ultimate goal of this book is to help unify the study of consciousness and attention: we believe that a focused examination into conscious attention will entail important theoretical progress that furthers our understanding of the human mind. To that end we present the major theoretical and empirical work in the related fields and define the key terms. Though it may be counterintuitive to unify the dialogue by focusing on the dissociation between consciousness and attention, we hope to convince you that attention is a predecessor to conscious experience and can operate independently from and outside of consciousness. The overlap between the two—conscious attention—is where it gets most interesting, because the functional purpose of conscious attention remains open to debate and there is a great deal of room for empirical work.

Acknowledgments

Our collaboration would have been impossible without the careful mentoring of Zenon Pylyshyn while we were at the Rutgers Center for Cognitive Science. We thank Zenon, members of the Visual Attention Lab, and those who attended our Object Group meetings for many thought-provoking discussions on the various aspects of attention. We would also like to thank Randy Gallistel, Fuat Balci, and Alvin Goldman for influencing the way we think about these topics. When a philosopher and a cognitive psychologist collaborate, many wonderful ideas are generated, but they may be written in strikingly different ways. We are very thankful to Victoria Frede for numerous critical suggestions that improved the book. A special thanks to Mary Rorty, to whom we owe the deepest debt of gratitude for her insightful, meticulous, and abundant comments and edits to the text—we are incredibly thankful for her patience, knowledge, and generosity. Finally, the exceptional editorial staff at the MIT Press deserves a mention, for the support from Philip Laughlin and Christopher Eyer, and especially for the careful edits we received from freelance copy editor Suzanne Schafer.

Now for some words from each of us individually.

Montemayor would like to thank the following friends and colleagues. Many of the philosophical ideas that inspired aspects of our theoretical proposals emerged from my discussions with Tony Bezsylko, Geoffrey Lee, Richard Price, James Stazicker, and Brad Thompson (the Mission philosophy of mind reading group). I am indebted to the insightful way of approaching philosophy that the members of this group displayed with incredible cheer at every meeting. I am also very grateful to the members of the Action and Consciousness Laboratory, particularly to Ezequiel Morsella and Allison Keiko Allen, for many discussions on the intricacies of approaching the distinction between awareness and self-awareness. I received invaluable feedback on some of the material on memory at the 2013 International Society for the Study of Time meeting in Crete, Greece. Finally, thanks to Rasmus G. Winther for his feedback on the evolutionary aspects of our proposal.

Haladjian would like to thank the following friends, family, and colleagues. This book is personally dedicated to my late father Vartkes Haladjian. It's not a spy novel, but it is about a mystery of sorts, so I know he would have appreciated it. I am especially grateful for the grounding in experimental psychology I received at Rutgers University, with wise guidance from my mentors Zenon and Randy. Some of the ideas for this book emerged in Anne Treisman's course on consciousness at Princeton University, and I feel especially lucky to have had the opportunity to discuss these issues with one of psychology's greatest and sincerest minds. The support from my mother, Georgette, my sister, Caroline, and my dearest friend, Chad Miller, has been invaluable and unwavering, as they all basically helped me survive graduate school. In Sydney, I would like to thank Alex Holcombe and Ahmed Moustafa for helpful discussions on attention and neuroscience. Notable moral support was provided by Myrna "D.J." Machuca-Sierra, Ella Wufong, and Kate Mayor, whose publishing advice was especially appreciated. And a special thanks to Blaine Arnold for the many lively late-night debates that challenged my empiricist bias and opened my mind to broader perspectives (and also for providing me with volumes of lossless digital music files to support the writing process). As much as I am driven to explain consciousness through empirical methods, there are still aspects of the topic that are difficult to grasp fully, akin to the concept of infinity. Nevertheless, I do have "faith" in science and in explaining theories through replicable independent observations, which probably will be proven wrong one day. Finally, this work was completed down under at the University of Western Sydney while I was on a Foundational Processes of Behavior postdoctoral research fellowship, and I am appreciative of the generous support that afforded me the time for the project.

1 Introduction

1.1 Theories of Consciousness, Attention, and Conscious Attention

Can there be a scientific theory of consciousness? An adequate scientific theory would have to satisfy the following desiderata: it must not only be parsimonious, empirically adequate, and logically consistent, but also maintain theoretical integrity. Recent studies in cognitive psychology and philosophy that offer theories of consciousness meet some of these criteria better than others. Further theoretical work is needed to assess which of the current theories is most likely to advance the study of consciousness as a scientific theory, to make it not only appealing but convincing.

When you first delve into this literature, many of the current theories on consciousness seem to be in opposition to each other. The incompatibility, however, is more apparent than real, because the resolution of underlying ambiguities may reveal the various candidates to be theories about different cognitive processes associated with consciousness rather than differing theories of consciousness itself. This book aims to offer an accurate and rigorous account of the most recent theories on consciousness in order to address theoretical gaps, with a focused consideration of how research on visual attention can provide the grounding for an empirically driven understanding of consciousness. As we will demonstrate, that goal is best served by including an evolutionary account of the development of cognitive systems (i.e., attention) into a theory of *consciousness and attention*.

The nature of phenomenal consciousness is a fundamental philosophical question that has received greater attention at least since the work of Descartes. It is a particularly challenging question to address because it is so intimately tied to the first-person point of view, and thus susceptible to different interpretations and difficult to study empirically. Whenever we refer to 'consciousness,'[1] (e.g., phenomenal consciousness or conscious awareness), we mean the qualitative character of an experience or mental state for a given subject at a given time, which is likely comprised of many cognitive processes. By 'attention,' we mean specifically the selective filtering and processing of sensory information, although attention is not restricted to perceptual processes. The cognitive mechanism of attention has often been compared to consciousness, since attention and consciousness appear to share similar qualities. But attention is defined functionally, whereas consciousness is generally defined in terms of its phenomenal character—what it is like to undergo a conscious experience, independently of functional considerations. This book offers new insights and proposals about how best to understand and study the relationship between consciousness and attention by examining their functional aspects. In so doing, we are ultimately led to the conclusion that consciousness and attention are largely dissociated.

Although the relationship between consciousness and attention has been at the center of recent debates about consciousness, the proposed accounts for such a relationship are based on problematic assumptions. For instance, some authors (e.g., J. J. Prinz 2012) assume that the terms 'attention' and 'consciousness' describe a single type of mental phenomenon and that, in principle, one can stipulate a list of cognitive processes from which to select such a phenomenon. It is unlikely that this approach will prove adequate, however, and one should avoid such assumptions for several reasons.

To give an example of the problems inherent in a reductionist (i.e., identity) approach, one can simply compare current theories for the different forms of consciousness and the different forms of attention. For instance, there is a consensus that there are different kinds of attention, with distinct neural correlates and functions that cannot be reduced to one another (see Parasuraman 1998). While many authors find it plausible to define 'attention' in terms of several basic forms of attention, other theorists think that defining it at all is hopeless (see Allport 1993, Johnston and Dark 1986). Similarly, some theorists think that there are at least two kinds of consciousness (Block 1995) and that only one of them is strictly related to awareness; others disagree (Dennett 2005). Many argue that the "hard problem" of consciousness makes its empirical study, unlike the study of perceptual attention, intractable (Chalmers 1996, Nagel 1974). Finally, there are theorists who think this "hard problem" of consciousness is just a pseudo-problem, and that consciousness can be and *must be* studied empirically (e.g., Churchland 1996, Dennett 1991). Since conscious experience is a process that occurs in the brain, it must have physical properties that can be explained empirically.

Because of the polysemy of the terms 'attention' and 'consciousness'—the divergence of opinion with respect to how they should be defined—it seems clear that one should avoid stipulating definitions without first delineating some theoretical and empirical constraints that such definitions must satisfy. Thus, we follow two guidelines in our inquiry into consciousness and attention. First, we base our proposals on considerations concerning evolution. There are well-known problems with accounts of the evolution of consciousness (Macphail 1998), but the situation is clearly different, and less problematic, with respect to the evolution of attention (e.g., Cosmides and Tooby 2013, Parasuraman 1998, Tooby and Cosmides 1995). We investigate current views on the evolution of different types of attention and explain their implications for the debate on the relationship between consciousness and attention. Second, considerations about evolution will help inform a broader understanding of a theory of *conscious attention*, in which we cover findings that are rarely integrated. In particular, we will explain the connection between conscious attention and the findings on 'flow' experiences associated with what some authors call effortless attention (see Bruya 2010). By conscious attention, we mean a phenomenal and reportable experience of perceptual information that is processed by selective cognitive mechanisms, and thus made available to higher-level cognition. In a philosophical sense, conscious attention requires a "demonstrative awareness that one is attending to that object," which also entails voluntarily maintaining attention to that perceptually selected object (Wu 2011).

In the service of conceptual clarity, this book analyzes recent theories on consciousness and attention by categorizing them in terms of a spectrum of theoretical complexity, from the theories that impose the strictest requirements on the interpretation of empirical findings, to those that allow the widest range of possible interpretations. Jesse Prinz (2012), for instance, has defended the view that consciousness is just attention. This identity view entails that there cannot be any finding about attention that is not also a finding about consciousness, and vice versa. Although this theory is parsimonious (it reduces the kinds of cognitive processes associated with consciousness and attention to a single type) it creates problems for empirical adequacy. Specifically, such a view would not account for findings that indicate some disassociations between attentional processing and conscious awareness. Should it not be possible that some form of attention exists without consciousness—for instance, in more "primitive" attentional systems—even if there is no possible form of consciousness without attention? Michael Cohen and colleagues (Cohen et al. 2012) argue in favor of this possibility. Toward the opposite end of the spectrum, one finds the view that consciousness and attention are often dissociated—that is, that there can be forms of consciousness without attention and vice versa. Such a position is advocated by Christof Koch and Naotsugu Tsuchiya (2007), among others. There is an even more extreme argument for full dissociation, according to which there is no overlap at all between consciousness and attention, implying that there is no such thing as conscious attention (Tallon-Baudry 2012 seems to favor this proposal).

One can argue that this wide range of theoretical perspectives stems from the lack of a common understanding of how to define 'attention.' Indeed, the other desiderata of logical consistency and general conceptual clarity require the disambiguation of fundamental terms. It is therefore crucial to determine whether or not the various theoretical perspectives are talking about the same type of attention, especially when the results suggest, for example, that there cannot be consciousness without attention. Are the proponents of differing views about consciousness and attention assuming the same type of consciousness? Are they assuming the same type of attention? These questions need to be addressed in order to specify the logical, inferential, and practical implications of the theories. To address these questions we present a rigorous survey of the current scientific literature. Based on the empirical findings, we propose definitions for forms of consciousness, forms of attention, and forms of conscious attention (see the glossary).

Our ultimate goal is to provide a systematic account of the relationship between consciousness and attention. The conceptual clarity that the spectrum of dissociation brings to debates on consciousness and attention helps achieve this goal. Throughout the book, we shall refer to our proposal concerning the dissociation between consciousness and attention as the *consciousness and attention dissociation*, or CAD. There are specific definitions of the kinds of consciousness and attention, and this clarification applies to all the alternative views. In terms of their level of dissociation (CAD), we categorize the spectrum of views we analyze as follows, ranging from the most restrictive views to less restrictive ones (see figure 1.1):

1. *Identity* Consciousness and attention are identical. This view reduces attempts to understand consciousness and attention into a single endeavor, and assumes that the empirical study of attention will inform our understanding of consciousness.

2. *Type-A CAD* There are several kinds of attention without consciousness, but only one kind of conscious attention. A form of attention may be necessary for consciousness, but consciousness is not necessary for attention. Exactly what kind of attention may be necessary for consciousness is an issue we explore throughout the book.

3. *Type-B CAD* All forms of consciousness are of the same kind, and there is a single kind of conscious attention; however, attention is not a necessary condition for consciousness, nor is consciousness considered a necessary condition for attention. This position would make empirical studies trickier to get right, since it is crucial to correctly identify cases where consciousness exists without attention and vice versa.

4. *Type-C CAD* The dissociation between consciousness and attention is greater, such that there are forms of attention without consciousness and *many* forms of conscious attention. The reason why this category of theory ultimately includes more dissociation is that in the absence of a unified form of conscious attention, the dissociations *within* conscious attention imply less unity in the relationship between consciousness and attention. Here again, attention is not necessary for consciousness, and vice versa; and again, it is tricky and crucial to correctly identify these cases empirically.

5. *Full Dissociation* Consciousness and attention are entirely separate, and explanatory models will produce two completely independent descriptions and mechanisms.

A primary aim of this book is to provide a meta-analysis of the findings on consciousness while enhancing the conceptual clarity of those findings. We hope to unify the way in which both philosophers and scientists refer to consciousness and its relationship to attention, since both fields are important to furthering our understanding of conscious awareness. We believe that thoughtful integration of the evolutionary aspects of consciousness and attention will serve to reshape the debate on this topic in a direction that favors less restrictive, more dissociative theories. Our main argument

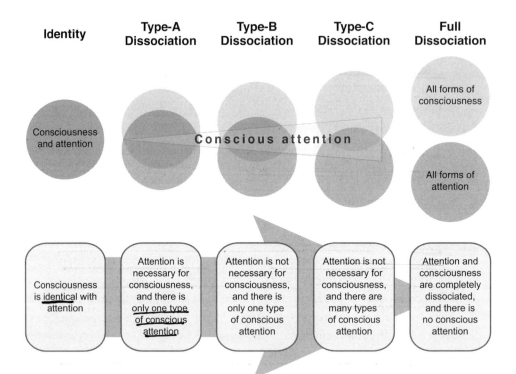

Figure 1.1
From left to right, the increasing degree of the Consciousness and Attention Dissociation (CAD), from identity to full dissociation. While it is likely that at least some form of crossmodal attention is necessary for consciousness, we argue that most views on consciousness actually entail greater dissociation.

regarding evolution is this: the thesis that attention is an adaptation—which is the most plausible thesis and receives unanimous support—is incompatible with most theories of the evolution of phenomenal consciousness. For this reason the spectrum of plausible views is significantly skewed toward a more dissociative interpretation. We see this as an indication that immediate progress in the debate is achievable.

In section 1.2 we present a historical overview of the development of the views described above, and then in section 1.3 we offer a more detailed argument about the connection between those views and the issues concerning evolution. This introductory chapter lays out the topics and structure of the book.

1.2 Problems of Consciousness: A Retrospect

Thomas Nagel (1974) famously said that "the problem of conscious experience is what makes the mind–body problem both interesting and intractable." The problem of how the mind connects with the world would lose its allure, and potentially become trivial or irrelevant, if one had no idea how solutions to this problem could explain consciousness. Once the current theoretical and scientific options to account for consciousness are carefully assessed, however, it becomes clear that they are all problematic. The clearest way of formulating the intractability of this problem is in terms of what David Chalmers (1995) called the "hard problem" of consciousness: Why would anything physical have the property of being conscious? That is, why should anything physical have conscious experiences? Directly related to this question is the problem of identifying the underlying physical brain processes that support the subjective experience of consciousness. Much has been written about this, and there is now widespread consensus that it is not only a challenging philosophical problem, but also one of science's biggest and most difficult unsolved puzzles. (Perhaps we can blame those stubborn zombies!)

Although the problem of consciousness is remarkably intricate, a great deal of progress has been made on the theoretical front. Significant conceptual clarity has been achieved with respect to the issue of why functions for cognitive processing may explain some of the forms of conscious integration required for working memory representations—what Ned Block (1995) calls *access consciousness*. Yet, these findings may not sufficiently account for the qualitative aspects of conscious experiences—what Block calls *phenomenal consciousness*. It is the qualitative aspects of our conscious experiences that are strictly related to the "hard problem" of consciousness. One topic we will address is how a theory of attention may help elucidate the relationship between access consciousness and phenomenal consciousness.

Another area of theoretical progress concerns debates about mental representation, content, and *qualia* (the qualitative aspects of individual conscious experiences). It seems that there may be changes in conscious experiences, or qualia, that cannot easily be reduced to the content of mental representations. For example, different experiences of color may represent the same type of color spectrum (as in 'inverted spectrum' cases).[2] Similarly, there seem to be cases, which Block (2010) calls 'mental paint,' in which changes in the voluntary focus of attention alter the content of experiences, without any corresponding changes in the stimuli themselves. It is an open question whether these are indeed legitimate aspects of phenomenal consciousness, but the theoretical framework for explaining conscious attention has been enriched by these definitions.

In order to understand common aspects of conscious experiences, recent research has focused on how voluntary and involuntary attention interact with changes in the content of experiences, or of experienced *contrasts*—that is, how attention interacts with changes in phenomenology, with respect both to the content of experience and to its degree of subjective effort. In cognitive psychology, ambiguous images are paradigmatic cases of such contrasts. In the Necker cube (a geometrically ambiguous image that appears to point upward or downward) or the 'duck-rabbit' drawing (a semantically ambiguous image that can look like a duck or a rabbit), although the stimulus itself does not change, the viewer experiences it in one of two alternative ways at a given time, but never both at the same time (see figure 1.2). An interesting issue is that although some attentional contrasts depend on the conceptual repertoire of the observer (as in the duck-rabbit), others seem to be independent of such content (as could be the case with the Necker cube and other geometrically ambiguous images).[3]

It is an established finding in vision science that such ambiguous images can alternate at a constant rate. At first, one interpretation is salient, then it recedes and the other, incompatible interpretation becomes the salient one. The observer, however, can also voluntarily direct her attention and flip the interpretations, for example, by focusing on one of the inner corners of the Necker cube (e.g., focusing on the lower inner corner where three edges meet as being the front face of the cube will cause the ambiguous drawing to be perceived as an upward-pointing cube). The observer consciously perceives only one of the images, and this image-dominance alternates over time. That is, the observer cannot see both interpretations simultaneously. These forms of contrasts in phenomenology and attention, however, seem to be dependent on a specific kind of ambiguous content that has multiple possible qualitative interpretations (e.g., a duck or a rabbit; a downward-pointing or an upward-pointing cube).

There are similar contrasts, however, that cannot be explained just in terms of content *ambiguity*. For instance, seeing a tree as either just a tree or as an elm has nothing to do with ambiguity. The two perceptual

Figure 1.2
The duck-rabbit and the Necker cube. When viewing the duck-rabbit image, one perceives it as a duck by focusing on the beak on the left side of the image; the rabbit can be perceived by focusing on the right side of the image where the rabbit's nose would be. Similarly with the Necker cube, one perceives an upward-tilted 3D cube by focusing on a corner in the lower middle area of the cube where three edges intersect; or perceives a downward-tilted cube by focusing on the upper intersection of the three edges. (This rabbit-and-duck drawing, "Kaninchen und Ente," appeared in the German magazine *Fliegende Blätter* in October 1892.)

experiences are different and yet the visual stimulus is the same, as in the case of ambiguous images. But here the interpretations do not alternate. This kind of case is one of *semantic specificity* rather than of interpretive ambiguity—that is, it is category membership that is at stake. Seeing a tree is different from seeing the same tree as an elm, and seeing a tree also differs from seeing it as an oak. Some of these changes in what one sees depend on voluntary attention, while others are more automatic and effortless.

One can make this point more precisely as follows. Seeing an elm *entails* seeing a tree, as one cannot have a visual experience of seeing an elm without having an experience of seeing a tree. Ambiguous images, like the duck-rabbit drawing, are different in this respect, since seeing a duck does not entail seeing a rabbit, and vice versa; indeed, seeing one image precludes seeing the other. These experiences depend on the Gestalt interpretation of the same perceptual information, often influenced by how attention is allocated. Changes that depend on expertise fall within this strictly conceptual phenomenon, and they tend to occur automatically, without much inferential or memory-retrieval effort. To take this problem into a different modality, experiencing a musical composition just as a melancholic melody differs from experiencing it as a melancholic melody that is a work by Beethoven. Many of the practical implications of what we experience hinge on this type of contrast and are related to our knowledge or expertise. The action of listening to sounds differs from the conscious activity of listening to Beethoven, and one can experience the activity either by effortfully monitoring aspects of the song, or by simply enjoying those components automatically, based on previous expertise.

How exactly such contrasts in content and attention should be understood is one of the issues we analyze in this book (see section 2.4 and chapter 4). A related problem is how to define the kind of experiential changes that are associated with what Block calls 'mental paint.'[4] In discussing the implications of a series of findings on visual attention by Marisa Carrasco and colleagues (e.g., Carrasco, Fuller, and Ling 2008, Carrasco, Ling, and Read 2004, Carrasco and Yeshurun 2009), Block argues that these changes in experience, based on changes in the allocation of attention, are not dependent on external changes in the stimuli themselves or on changes in conceptual aspects of the stimuli related to semantic ambiguity or expertise. The findings from those studies demonstrate that instead, changes in attention alter the features of unchanging stimuli. In a variety of experimental conditions, a simple voluntary shift of attention to one stimulus from another can alter the perceived contrast, size, and other features of the stimuli involved—a change that is fundamentally dependent on the allocation of attention. Block claims that the quality of these experiences feels unreal, similar to visual experiences such as afterimages, and calls the experiences forms of subjective unreality. So from a strictly empirical point of view, we need a comprehensive theory of attention that draws the distinctions documented by experiments on attentional shifts, Gestalt reversals, and the typical forms of attentional deployment (e.g., top-down and bottom-up, voluntary and involuntary).

The reason the findings discussed by Block are puzzling is because they demonstrate that shifts in the focus of attention alter consciously experienced features of unchanging perceptual stimuli. That is, the perceptual input remains the same, but the experience of it changes. Block contends that these findings cannot be explained as illusions because the percept relies on how attention is allocated rather than being a misrepresentation of the stimuli. Their subjective unreality, Block claims (2010, 54), has not received any empirical investigation. They also remain unaccounted for in a broader theoretical treatment of consciousness (i.e., conscious attention). We argue in chapter 3 that the phenomenon of 'mental paint' is an important factor in the dissociation between consciousness and some forms of attention and explain in chapter 4 that these shifts characterize the overlap between phenomenal consciousness and voluntary attention.

Another source of problems afflicting treatments of conscious perception is the role of the self in phenomenal experience. A number of intricate questions originate from this topic. Can one be conscious of something, whether an emotion or a perceptual representation, without also being conscious that one is conscious of it? Is the self, as Immanuel Kant (1929, 362–3) argued, a necessary condition for every possible experience without itself being experienced? How should we understand consciousness, self-awareness, and the conscious self? So-called higher-order theories distinguish between experiences and conscious experiences (see Carruthers 2000, Lycan 1996, Rosenthal 2002). The distinction is generally made by appealing to the privileged point of view of the subject of experience, and exploring what it is like for a subject to have a particular experience. This could be interpreted as a necessary condition of minimal subjectivity for conscious experiences: the point of view of the subject of experience (see Merker

2007a). For analogous reasons, some authors claim that the self (in some sense) is indeed constitutive of conscious experiences (Damasio 1994).

One problem with an emphasis on the self, however, is that it seems to demand too much of conscious creatures and, for that reason, it also seems to be empirically implausible. Koch (2012a), for instance, argues that the notion of self is not necessary for conscious experiences. That is, one can have consciousness without having any representation of oneself. He criticizes the 'mirror test,' which infants and most animals fail, as a test for consciousness, even though it seems to be a good test for self-awareness. The reasoning is that infants and many animals must have some kind of phenomenal consciousness because they experience pain, emotions, and other bodily sensations. They may not have self-consciousness but, the claim is, they do have phenomenal consciousness.

Plausible as this criticism is, however, the relationship between consciousness and self is much more intricate. In a passage in which Koch is defining the scientific problem of consciousness (what he takes to be the scientific issue, the main topic of his 2012 book), he uses two incompatible interpretations of the word *self*, specified below by our use of underlined italics and bold fonts. Koch is criticizing the conclusion that failure to pass the mirror test indicates the lack of consciousness:

> One among many observations makes this conclusion implausible. When *you* are *truly engaged* with the world, *you* are only dimly aware of **yourself**. *I feel* this most acutely when I climb mountains, cliffs, and desert towers. On the high crag, life is at its most intense. On good days, *I experience* what the psychologist Mihaly Csikszentmihalyi calls *flow*. It is a powerful state in which *I am exquisitely conscious* of my surroundings, the texture of the granite beneath my fingers, the wind blowing in my hair, the sun's rays striking my back, and, always, always, the distance to the last hold below me. Flow goes hand-in-hand with smooth and fluid movements, a seamless integration of sensing and acting. All attention is on the task at hand: The passage of time slows down, and **the sense of self** disappears. That inner voice, my **personal critic** who is always ready to remind *me of my* failings, is mute. Flow is a rapturous state related to the mind-set of a Buddhist lost in deep meditation. (Koch 2012a, 37, emphasis added)

He later continues: "This loss of **self-awareness** not only occurs in alpinism but also when making love, engaging in a heated debate, swing dancing, or racing a motorcycle. In these situations, *you are in the here and now*. You are in the world and of the world, with little **awareness of yourself**" (Koch 2012a, 37). It is clear that the 'self' (highlighted in bold font) is the higher-order self that a subject recognizes in claiming a particular thought as hers and in thinking of herself as herself. We can call this the *recognitional self*. Koch seems to be justified in claiming that the recognitional capacities associated with this kind of self may not be necessary for consciousness. Surely infants are conscious, so does it matter if they cannot pass the mirror test? But how can we identify the 'self' as highlighted in italics—the *phenomenal self*? This is a central question that must be answered in order to address the relationship between higher forms of self-awareness and phenomenal consciousness. Later in this book, we offer insights that can help elucidate such a relationship by analyzing the relationship between consciousness and attention.[5]

The foregoing discussion illustrates that although there has been progress in recent years on the theoretical front in consciousness research, a lot more needs to be done. Specifying the most urgent issues in the theoretical agenda (in chapter 4) and proposing ways to better understand conscious content (in chapter 5) are other key goals of this book.

In addition to the theoretical progress on consciousness, remarkable progress has been made on the experimental front in consciousness research. Although it remains challenging for investigators, the empirical study of consciousness—once virtually ignored—now makes use of substantial laboratory resources and produces valuable evidence related to the nature of conscious awareness.

What happened to encourage such a shift? Interestingly, the way toward progress in the empirical study of consciousness was paved by experimental research on unconscious perception and, more generally, unconscious cognitive processing. A very influential and productive approach, proposed by Bernard Baars (1988), is to compare conscious processing with well-known unconscious processes studied in cognitive psychology using established research methodologies. Such research includes studies on the implicit processing of a visual stimulus when it is masked, binocular rivalry (where different visual information is presented to each eye so that the conflicting information competes for awareness), and change blindness (the inability to detect a significant change in a visual scene). We will discuss these instances in more detail in chapter 2. The comparison between the neural correlates of conscious and unconscious processing has already produced important insights into the nature of conscious awareness. For example, the thesis that consciousness is the result of a highly integrative process has been confirmed with neuroscientific evidence (e.g., Dehaene and Naccache 2001, Di Lollo, Enns, and Rensink 2000). There are also models that aim at implementing the computational underpinnings of such highly integrative processes in the light of the theoretical issues discussed above (see Tononi 2004, 2008, 2012). This integrative theory is something we will consider in more detail later in the book.

Another major area of growth is the range of topics that are studied experimentally in hope of achieving a better understanding of consciousness. For example, researchers are exploring consciousness in dreams and daydreaming, and its relation to the default neural network (i.e., the neural activations of the brain during states of wakeful rest). Experiments on conscious and unconscious perception, hallucinations, and mental imagery have also expanded our knowledge of perceptual conscious awareness. Research on the distinction between conscious decisions for action and the unconscious processes that guide motor control has shown that the processes reaching conscious awareness are indeed just the tip of the cognitive-processing iceberg (see Rosenbaum 2002).

As is the case with the progress on the theoretical front, major problems still loom on the horizon for an empirical approach to understanding consciousness. One of them is how to identify and explain all the nuances of our theoretical understanding of consciousness at the neural level. For example, even if experimental evidence confirms that consciousness correlates with a specific pattern of neural activation, what would that finding signify? Would it allow us to distinguish access consciousness from phenomenal consciousness? Could it be that the pattern of activation is literally just correlated with consciousness, but neither explains nor identifies what is truly unique about it? Much has been said about this issue, and we do not take a stand about the relevance (or irrelevance) of attempts to identify the neural correlates of consciousness, although we will comment on the main views. Despite the difficulties underlying the metaphysics of consciousness, we believe that the progress on the experimental front has been substantial.

There are other books, many by some of the leading researchers mentioned above, that cover in great detail these areas of theoretical and empirical progress, as well as the challenges that must be met.[6] Here we shall concern ourselves only with how the progress on all these topics relates to the largely unexplored issue of conscious attention. Thus, although we will canvass the main views in the literature and critically assess the theoretical and empirical discoveries, our main focus is the relationship between consciousness and attention. We believe that a clear account of this relationship is a necessary condition for an adequate theory of consciousness that can successfully guide empirical research.

In another recent theoretical development, philosophers have started studying visual attention with the goal of arriving at a definition of attention that specifies the relationship between consciousness and attention (see Mole, Smithies, and Wu 2011). The exact nature of this relationship remains an open question, and the task of specifying it presents many challenges that have not been properly acknowledged in the literature. We propose that taking evolution into account will help clarify the relationship between consciousness and attention. For example, we will examine whether or not both processes evolved at the same time. In spite of the recent optimism about views that identify consciousness with attention, evolutionary considerations complicate their relationship.

1.3 The Relationship between Consciousness and Attention in the Light of Evolution

One way of addressing the conundrum of the relation between consciousness and attention is by comparing it to successful reductions, like the reduction of questions about life to questions about DNA. This starts with a critical assessment of the *identity thesis* for consciousness and attention. According to this view, consciousness just *is* attention (see J. J. Prinz 2012). There are advantages to this view, which we will discuss later. But there are also major problems, both theoretical (see Koch and Tsuchiya 2007) and empirical (see Kentridge 2011, Tallon-Baudry 2012).

Many of these problems are best understood as possible responses to two different questions. (1) Are all forms of attention forms of conscious attention? The intuitive response is yes—anything that is attended (i.e., filtered selectively by sensory processing mechanisms) must be a form of conscious attention. When one looks at the empirical evidence, however, things are not so clear-cut; many forms of attention never reach conscious awareness. (2) Conversely, are all forms of consciousness forms of conscious attention? That is, does everything that enters conscious awareness indicate the presence of conscious attention? Here things are even trickier, and no obvious response seems without problems. The leading intuitions come in various epistemic or metaphysical flavors, but none of them clearly commands the field. Furthermore, this is true not only of theoretical issues: when one looks at the empirical evidence, things are equally problematic.

The rejection of the identity thesis seems inevitable, and will allow a wide landscape of options to emerge. There seems to be attention without consciousness, as in the case of blindsight, among other cases (see section 2.3). Then the question is how prevalent these forms of unconscious attention are, and to what extent they guide cognitive processing. There also may be consciousness without attention (although this is more difficult to identify), and the same consideration about scope is pertinent to that idea. Depending on the degree of the dissociation between consciousness and attention, one can envision several possibilities with different critical theoretical implications.

What are the possible outcomes? Suppose the degree of dissociation is insignificant. In that case, one could distinguish a few forms of consciousness without attention (or vice versa), but they would be very rare cases of little consequence, such that one could almost identify consciousness with attention (we could classify this as a Type-A CAD). Even in this case, however, several questions emerge. Why would any form of consciousness without attention (or vice versa) exist? Assuming that attention is an adaptation, why would some forms of attention emerge before others? How should we reconcile the evolution of attention with the notoriously difficult question about the evolution of conscious awareness?

Issues of scope are relevant here. Suppose that 'conscious attention' refers to a single kind of mental phenomenon, but that there are a few cases of consciousness without attention, or of attention without consciousness. This supposition, involving Type-B CAD, would suggest that those forms of consciousness may not have been easily integrated with attentional processes, a surmise that could point to distinct stages in the evolution of consciousness and attention—some forms of attention being more primitive than others, and some more resilient than others to integration with other cognitive processes. Or perhaps it has nothing to do with evolution, but with two fundamental kinds of consciousness, one essentially linked to attention and the other independent from it. This is another intriguing possibility that would need elaboration.

Suppose, on the contrary, that the degree of dissociation is severe (Type-C CAD). Some cases of conscious attention could be highly representational and dependent on external sensory information for mental content (as attention is generally understood); while other cases of conscious attention could be directed to the conscious self. And of course, there will be cases in which attention is not accompanied by consciousness at all, or at least not by phenomenal consciousness. There may also be cases in which consciousness is not accompanied by attention, such as, for example, a pain that has been occurring without being noticed, or the constant sound of an air-conditioner in the background that has been unattended (but see the description of the 'cocktail party effect' in section 2.1.2). The main result would be that consciousness and attention rely on two fundamentally distinct processes, which may be integrated in some cases, but which are largely independent from one another.

If the hypothesis of severe dissociation, which we believe is a more plausible view than the identity thesis, were to be accepted, then questions about evolution would become fundamental to a theory of consciousness. For that would suggest that different forms of attention evolved at different times, and also that some forms of consciousness evolved at different times. It would also be plausible to think that different forms of conscious attention evolved at different times and that they eventually became integrated in the manner of a "kluge" (see Marcus 2008). Thus a variety of evolutionary designs would ground the cognitive processes that we generically label as 'consciousness' and 'attention.' These labels would not identify what philosophers call natural kinds, but rather would point to a variety of fundamentally different cognitive processes. Primitive forms of consciousness would differ in purpose from more recent and integrated forms of conscious attention. Such primitive forms of consciousness could presumably be found in many animals, and here the question would be how many species would have these primitive forms of consciousness, and whether these forms of consciousness are dissociated from attention. Another challenge would be to explain how severe dissociations, and the diversity of the forms of consciousness, are compatible with the *unity* of conscious experience, which is generally how we experience the contents of consciousness: as a highly unified and integrated total experience, rather than a sum of discrete ones.

These considerations about the evolution of attention and consciousness and how they are related are central to the contribution we hope to make with this book. Although conscious attention seems to be a highly unified process, evolutionary considerations strongly suggest that different types of conscious attention—for instance, effortful versus effortless voluntary attention—are independent from each other and may have evolved at different stages. So one conclusion we draw is that the identity thesis, despite its intuitive appeal, is too simplistic to account for the evolution of both consciousness and attention. Considerations about evolution, which we frame within our proposed types of dissociation, constitute one among many approaches we will bring to bear on this problem.

To further make our point: it is uncontroversial that different areas of the human brain evolved at different times (e.g., see Striedter 2005). It seems that the cerebellum and other older areas of the brain are not necessary for consciousness (Koch 2012a, Tononi 2004); and yet the cerebellum is crucial for navigation, and thus has several areas devoted to "attending" to features of the environment, albeit in a sense that needs to be specified. Areas associated with emotion, perception, and motivation, mental attributes that were thought to be deeply related to phenomenal consciousness, are also unnecessary for conscious awareness. So, based on neuroscientific findings, one can make a very plausible case for dissociation, as we will discuss in section 2.2 and in chapter 5. This conclusion has to be evaluated in conjunction with the considerations that led theorists to propose the identity thesis. This is one reason we believe it is important to provide an integrated account of consciousness and attention based on the latest psychological and neurological findings. By doing so, we can help elucidate theoretical distinctions that are fundamental for a better understanding of the conscious mind, such as the distinction between those higher forms of self-awareness associated with the recognitional self and the more minimal forms associated with the phenomenal self.

Attention research in cognitive psychology presents an even bigger challenge for integrated accounts such as the one we pursue here. Most psychologists working on attention, because of the intractability of the problem of consciousness, had either no interest in consciousness or no way to connect their findings with considerations about consciousness. Findings on change blindness, inattentional blindness, object tracking, and other aspects of attention shaped our understanding of visual attention without explicitly resolving the question of whether they were compatible with theories of consciousness. Making these connections explicit is another area in which we hope to contribute to the understanding of conscious attention. One of the main insights from research findings on attention, and one that will guide our inquiry, is that attention is concerned with connecting cognitive processing with objects in the external world, and seems more analytic and selective in nature than consciousness. Consciousness seems highly integrative, but its purpose is often questioned. For instance, some studies have found that the execution of the motor commands to perform an action occur before the "conscious decision" to make that action. Why, then, do we even need this subjective feeling of consciously making a decision?

With respect to subjectivity, we shall make a few clarifications about how the degree of dissociation of consciousness and attention applies to theories of consciousness that appeal to creature consciousness, self-awareness, or the first-person perspective. Throughout this book, our main focus is how the degree of dissociation between consciousness and attention (CAD) applies to any account of *state consciousness*—that is, to the views according to which being conscious equals being in a particular experiential state. This choice of theoretical focus is based on considerations that aim to maintain clarity and precision in our discussion. A theory of consciousness that appeals primarily to creature consciousness, self-awareness, or the first-person perspective would introduce the ambiguities and controversies of the debates concerning subjectivity. This is not to deny the plausibility or importance of such views. Rather, we believe that by showing how CAD applies to different versions of the 'state' view, its application to 'creature' or subjectivist views will become clear.

Although we focus on state consciousness views, we do not shy away from the challenging topic of subjectivity. We explain at critical points why subjectivist approaches actually *increase* the degree of dissociation between consciousness and attention, favoring a higher level of CAD. We clarify how different forms of attention may relate to different forms of subjectivity, and we shall mention two critical issues that demonstrate why considerations about subjectivity lead us to posit a more severe dissociation between consciousness and attention.

One critical issue is the nature of the self, or first-person point of view. Although a Cartesian dualism is not a popular option (for good reasons, we believe), some authors think that there has to be a primitive metaphysical property associated with having a first-person point of view (see Baker 2013, Hare 2007, 2009). If some kind of primitive, metaphysical perspective must be assumed, then the spectrum of dissociation between creatures with such a perspective (according to some, only humans; see Baker 2013) and creatures without it but *with* empirically confirmed capacities to attend to features of the environment is rather severe. This implies an extreme type of CAD, if not full dissociation.

A second problem is that some authors argue that linguistic capacities are necessary to develop a sufficiently robust first-person perspective—one that makes autobiographical memory, self-reference, and metacognitive states possible. More specifically, the first-person perspective seems to necessarily involve indexicals (i.e., explicit reference to the subject of experience as oneself, which requires mastery of the concept 'I'), along with all their implications for action, thought, and memory. These are plausible views, although there is much controversy regarding indexicals, ranging from views that strongly deny the need for first-personal indexical content in order to explain action (e.g., Cappelen and Dever 2013, Millikan 1990, 2001) to views that hold that linguistic capacities for essentially indexical content are indispensable for a first-person perspective (Baker 2013, Metzinger 2003). It is clear, however, that making language capacity a necessary condition for subjectivity makes the level of dissociation much more severe than if one only focuses on mental states. If linguistic capacities were necessary in order to have conscious awareness with a strong subjective component, then the dissociation between such conscious awareness and most forms of attention would also be quite severe. (We explain why similar considerations apply to 'field' views, based on structural considerations, in chapter 3.) Thus, another advantage of our approach is that we demonstrate the implausibility of the identity view by using the account that makes it most plausible: the state view.

In the end, we believe that an adequate theory of consciousness can be achieved only by engaging in an integrative effort that contrasts the recent findings on attention with the recent findings on consciousness, in the light of specific desiderata that must include biological plausibility and should be informed by the theory of evolution. We hope that such an effort, undertaken in the present work, will further improve our understanding of consciousness, attention, and conscious attention while illuminating the dialogue between philosophers and scientists. To date, no theory of the evolution of conscious attention has been carefully articulated. Filling this theoretical gap is one of our major goals.

1.4 The Spectrum of Dissociation at Work

The framework we provide will help to reshape the debate on the relationship between consciousness and attention. We will soon illustrate how this framework reinterprets theoretical approaches to that debate (e.g., higher-order theories versus first-order theories; phenomenalist versus representationalist theories, etc.). Although significant clarity emerges when debates are reinterpreted in this way, the framework is not entirely neutral. Its basis on empirical considerations concerning neural correlates and evolution favors specific forms of dissociation.

The present framework is the first to provide an overall assessment of consciousness *and* attention. Unlike other proposals for unifying the disparate views on consciousness or attention, our spectrum-of-dissociation approach will constrain, both empirically and theoretically, current views on consciousness and attention and, we hope, produce more meaningful exchanges with respect to their relationship. Guidance on a topic that is so difficult to approach requires some normative components—statements that establish how we *should* understand the relationship between consciousness and attention—in addition to strictly descriptive ones.

To illustrate this point: consider Francisco Varela's (1996) characterization of the debates on consciousness.[7] According to Varela, there are two axes with four theoretical directions, which map out all the possible views one may have about consciousness (see figure 1.3). Varela's formulation contrasts phenomenology to reductionism on one axis, essentially pitting subjective experience against objective measures. The other axis contrasts functionalism to mysterianism, which is the idea that consciousness cannot be defined functionally. Consciousness is ultimately defined by reaching compromises between these two axes of incompatible views.

While this chart is a useful way to convey graphically a range of the extant views on the nature of consciousness, it has some disadvantages. First, as will be argued in chapter 3, it may be too simplistic to categorize views on consciousness by opposing phenomenological views to reductionist views and mysterian views to functionalist views. But, more important, Varela's goal in providing this characterization of possible views was strictly

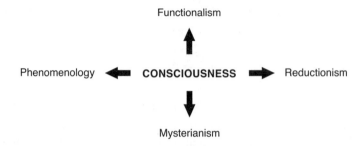

Figure 1.3
Possible views on consciousness can be charted within this two-dimensional categorization scheme. Adapted from Varela (1996).

descriptive—it was meant to classify extant views without introducing any distinctions more subtle than already allowed by those views.

In contrast, the spectrum of dissociation that we develop presents a much more finely grained set of possibilities, more intricate than any extant view is capable of capturing. In fact, introducing a spectrum of dissociation, assuming one accepts its general outlines, would substantially complicate Varela's chart. For example, one can be a mysterian about consciousness and hold an identity thesis, so that one ends up being a mysterian about attention as well. Conversely, one could be a mysterian but hold a full dissociation view, assuming thereby that attention can be fully captured by functional and reductive approaches; and so on with other views and levels of dissociation. Thus, the number of options grows significantly. Crucially, the spectrum of dissociation we propose makes a normative point, because it does not seek to capture the options available according to extant theories. Rather, it shows that these theories should be reconceived, and that the number of options is much larger than we had previously thought. This is why we say that the spectrum of dissociation makes a normative point rather than a merely descriptive one.

Moreover, there is another sense in which an analysis of the spectrum of dissociation is normative. Our use of evolution as a way of reframing the debate on consciousness is not neutral. It aims to find an empirical, biologically inspired framework for the study of consciousness, one that has been applied to other cognitive processes (e.g., memory, attention, navigational abilities). By demonstrating the superiority of dissociation views, our approach seeks to establish unambiguous guidelines for future research,

Introduction

which will be both theoretically based, on a uniform account of 'consciousness' and 'attention,' and empirically based, on the findings available to date.

1.5 Summary

To summarize our goals, the proposed spectrum of dissociation, or CAD, will be informed by empirical considerations that establish how any claims about consciousness and attention, as well as their relationship, should be addressed. It does not merely describe a spectrum of *possible* views one may hold—though it performs that task adequately—but it also imposes normative criteria with respect to how best to understand current behavioral and neural research on the correlates of consciousness, attention, and conscious attention.

The project of this book requires a review of the current theories and findings in attention and consciousness before we can address conscious attention. Chapter 2 focuses on different forms of attention that closely interact with each other, but which are characterized by fundamentally different functions and seem to have varying levels of dissociation from each other. Chapter 3 presents a theoretical overview of the literature on consciousness, and argues that extant accounts of the nature of consciousness entail some type of dissociation, most entailing a major form of dissociation such as those we characterize as Type-B or Type-C. Chapter 4 explains the nature of the types of conscious attention, emphasizing the claim that a high degree of cognitive integration is necessary, but yet may not be sufficient, for conscious attention. Finally, chapter 5 presents our case for the dissociation of consciousness and attention based on evolutionary considerations. While we believe that attention (perhaps of the crossmodal kind) is necessary for consciousness, our principal goal is to argue that dissociation views must be favored, even if the most plausible view turns out to be a Type-A dissociation view.

2 Forms of Attention

When the topic of attention is introduced in a lecture or a written work, the philosopher and psychologist William James is often quoted. So we shall adhere to this tradition and quote James's description of attention as the ability to focus on "one out of what seem several simultaneously possible objects or trains of thought," a process that requires "the withdrawal from some things in order to deal effectively with others" (James 1890/1905, 403–4). This intuitive and perhaps too simplistic description of the selective processing of perceptual and cognitive information has motivated a thriving area of research, particularly in the last 50 years.

Our discussion in this chapter focuses on *visual attention*, which has been studied extensively in cognitive psychology and neuroscience (although the findings are generally applicable to the other sensory modalities). A comprehensive review of this field would require several volumes, and specialized reviews on the various forms of attention have been done nicely by many others (e.g., see Bundesen 2001, Carrasco 2011, Driver 2001, Friedenberg 2013, Parasuraman 1998, Posner 1980, Scholl 2001, Theeuwes 2010, Wright and Ward 2008, Wu 2014); so we will try to limit our discussion to the major empirical findings on visual attention that have implications for consciousness. This review is important for our argument that attention and consciousness must be dissociated at some level, since there are functionally different types of attention that seem to operate independently and to have evolved at different times.

Attention, broadly defined, is a selective processing mechanism (or rather, a group of mechanisms) that enhances and selects perceptual information for executing actions and higher-level cognition, and, as we will argue, for consciousness. What is selected is often what is most visually salient; but attention can also be willfully directed toward a specific area or feature to guide the selection process, as, for instance, when looking for your red pen on a desk cluttered with multicolored objects. In a recent book, Christopher Mole (2011) argues that we should think of attention as a *description* of how something is done—absently or attentively—instead of trying to identify the specific mechanism that produces attention. This functional approach frames our problem nicely, since, as you will see, the many components of attention cannot be described by a single mechanism, but rather constitute a collection of cognitive processes that produce the ability to be attentive, that is, the ability to process selective information. From the start, we are able to identify dissociations among the different forms of attention and the brain structures supporting them, which will impact how we view their relationship to consciousness.

Attention is a central aspect of perception, and the manner in which attention is directed can dictate how visual information is perceived, often resulting in different interpretations of a single perceptual stimulus. Such 'phenomenal contrasts' are exemplified by the Necker cube or the duck-rabbit drawing we instanced in figure 1.2. These images can produce alternating phenomenal experiences automatically, regardless of the intentions of the observer; however, the observer can also direct her attention to certain features and thus change the interpretation. Such examples illustrate attention's important role in determining how a stimulus is perceived when there are multiple possible interpretations of the same perceptual input. Attention, therefore, is not only important for the selection of information for conscious awareness, but also important for how that information is interpreted.[1]

To better understand the role of attention in phenomenal experience and consciousness, we need to identify the many ways it has been classified and studied over the years. These theories generally have been based on what features, locations, or objects attention is directed toward, or on the manner in which attention is deployed, whether automatically or intentionally. In this chapter we describe these various aspects of attention, to give a better sense of how it may be related to consciousness. We begin in section 2.1 by outlining the 'varieties' of attention and then differentiate how these varieties are controlled, whether by 'bottom-up' or 'top-down' processes. In section 2.2 we describe the brain systems that have evolved to support these varieties of attention. This will help clarify the dissociations among the forms of attention, which has implications for the general dissociation between attention and consciousness described in section 2.3. The relationship between consciousness and attention, or conscious attention, will be discussed in section 2.4, although it will be examined in more detail in chapter 4. We conclude this chapter with section 2.5 by describing open questions regarding the empirical study of attention and its relation to consciousness.

2.1 General Classifications

As the quote from James suggests, it seems obvious what attention is. You can attend to something and notice it, or not; and this should indicate whether attentional mechanisms were implemented or not. Although the phenomenal experience of attention can be described in this simplistic way, the empirical study of attention is not so straightforward. Attention can be conceptualized in various ways, and several distinct attentional systems have been identified. In addition, there are also pesky, hard-to-measure forms of attention, like the 'preattentive' ones, that do not clearly correlate with what is experienced in awareness, and yet are crucial for behavior, perception, and conscious attention. Such nuances complicate our work even further.

As the empirical study of attention grew in the twentieth century, it became generally accepted that attention is comprised of several different cognitive processes with distinct neural structures. There are systems that can maintain alertness, or a state of vigilance in a task; there are orienting systems that prioritize certain sensory input via a specific modality or location; and there are higher-level executive functions that assist target detection, recognition, and awareness. Attention also can operate among several levels of complexity, with preattentive, low-level, and high-level stages that include both distributed and focused attention. To describe the automatic low-level processes and the more deliberate and focused forms of attention, a common distinction conceptualizes attention as 'bottom-up' versus 'top-down' processes respectively. Similarly, attention can be thought of as effortless and almost automatic, or it can be effortful and guided by a feeling of willful selection.

A common problem with these many forms of attention is confusion about the use of certain terms and what they actually imply, especially in the context of studying consciousness. One of our goals is to distinguish among and clarify the meaning of these forms of attention, how they relate to each other, and how they relate to consciousness.

2.1.1 Varieties of Attention

The main forms of visual attention we will discuss are those concerned with processing features, spatial information, and visual objects. This classification, along with the way in which they are modulated (i.e., bottom-up versus top-down), is a useful way to organize the primary functions of attentive processes. We present only an overview of the vast research that has been published in these areas, but within each of the following subsections we will suggest resources for more in-depth reviews.

Feature-Based Attention Attention can be drawn to certain features in a visual scene, and this *feature-based attention* is closely tied to bottom-up, low-level selection processes that are modular in nature (in theory similar to the modularity proposal by Fodor 1983). In the case of visual attention, the modular systems refer to specialized receptors and processing systems in vision that detect various kinds of visual information such as color, motion, or segment orientation. Extensive research on the visual cortex indicates that it is organized according to such specializations. Feature-based attention interacts with this organization and generally corresponds to the modular processing of different features by the visual system (for reviews, see Maunsell and Treue 2006, Theeuwes 2013, Treisman 1998).

An important role of feature-based attention is the ability to highlight certain features for selection, as in a 'visual search' task (Wolfe 1994). For example, if you are looking for your red pen on a cluttered desk, attentional systems can selectively bias color processing toward the detection of red-colored things to facilitate their selection. In this way, feature-based attention can be modulated toward features in a visual scene to increase the likelihood that they will be selected for further processing (e.g., Bacon and Egeth 1997). This biasing may be influenced by a willful (top-down) form of attention, but there are good reasons to believe that low-level factors are more likely to bias feature-based attention. For example, it seems that this type of attention is sensitive to context and priming effects. Being cued with the actual visual feature that must be located, as in a lab-based visual search task, facilitates feature detection better than simply trying to look for the target using top-down direction (see Theeuwes 2013).

Forms of Attention

When features are detected instantly during a search task, attentional processes are described as showing a *pop-out effect* (see Duncan and Humphreys 1989, Theeuwes 1992, Treisman and Gormican 1988). Pop-out is considered an early selection process that grabs attention based on the distinct saliency of a feature. For instance, something bright red among gray objects will pop out, resulting from bottom-up parallel processes and local comparisons of feature information. Attention is captured quickly and effortlessly, but the efficiency will depend on various factors, including how different this object is from surrounding distractors, and can be enhanced with multimodal cues (Theeuwes 1992, Theeuwes, Kramer, and Atchley 1999, Van der Burg et al. 2008). See figure 2.1 for some examples of pop-out.

The detection and processing of basic features is a low-level process, but the ability to report on them requires some sort of focused attention. *Feature*

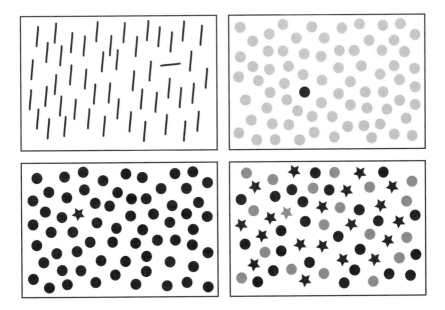

Figure 2.1
Examples of stimuli used to demonstrate the pop-out effect. Distinct items are easier and faster to detect when they are salient or sufficiently different from their surroundings, regardless of set size. The panel on the bottom right depicts a more difficult test array where the target (the grey star) is harder to find since it has a conjunction of features (feature conjunctions are notorious for slowing visual search and reducing the pop-out effect).

integration theory, proposed by Anne Treisman and colleagues, describes the manner in which attention processes and accesses features (Treisman 1988, 1993, Treisman and Gelade 1980). According to this theory, the creation of detailed object representations first requires the features to be processed through an early 'preattentive' stage, as described by initial theories of attention (e.g., Broadbent 1977, Neisser 1967). As this information is processed, it activates feature maps, which are separate cortical "spaces" for different types of features (e.g., shape, color, or segment orientation). After this preattentive stage, focused attention is required to bind and maintain these features together so that a representation of the object can be stored in visual working memory. (See figure 2.2 for more information on feature integration theory.) This is especially important for the detection of targets with conjunctions of features, since searching for a target object with several features requires the correct binding of those features. Consequently, this model aims to address the *binding problem*, which is the problem of how several features can be integrated to form a coherent object representation (for reviews, see Holcombe 2010, Treisman 1996, 2006, Wolfe and Cave 1999). We will discuss feature binding and object-based representations shortly.

In support of a modular organization of feature-based attention, neural studies have identified a hierarchically organized visual cortex, with basic feature processing occurring in lower and evolutionarily older areas of the brain.[2] After visual information from the retina reaches the primary visual cortex, it travels via two streams, a dorsal and a ventral stream, which have been described as the 'where' and the 'what' pathways respectively (see Ungerleider and Haxby 1994). Thus the dorsal pathway processes location and motion information, while other features, like color or shape, move up the ventral stream. Another characterization of these two pathways is that the dorsal stream supports 'vision for action' while the ventral stream supports 'vision for perception' (see Milner and Goodale 2008, Westwood and Goodale 2011). Regardless of how one chooses to characterize these two streams of information, it is important to note that some visual information may not overlap between the two pathways. This distinct channeling can produce sources of dissociation between the information provided for detecting object locations and for recognizing objects, or dissociations between the information for perception and that used to perform actions. (More on these processing pathways and possible dissociations in section 2.2.)

Forms of Attention

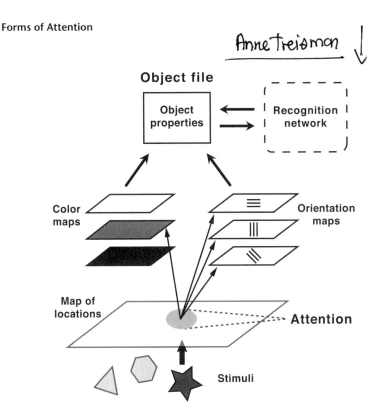

Figure 2.2
The components of feature integration theory, which demonstrates aspects of feature-based attention and its relation to spatial attention. Different systems process different features, such as color and orientation, with their own feature maps. These feature maps have a spatial component, such that the activation of a particular feature also identifies a location. There are also interactions between the different systems to enhance or inhibit attentional processing. Importantly, when serial attention is directed to an object occupying a particular location, the corresponding neurons in the various feature maps are activated such that those active features can be bound together as belonging to the same object (based on spatial coordinates). Such bindings are thought to form 'object file' representations. These representations can influence or be influenced by items stored in long-term memory, or the 'recognition network.' This figure is based on Treisman (1998).

Overall, feature-based attention is a fundamental and basic form of attention upon which richer representations are built. In fact, one can think of these bits of information as the building blocks of mental representations and visual consciousness, even though they need not reach conscious awareness to influence behavior. This form of attention can be described as a more primitive type of attention, operating almost automatically, based on early adaptations for the processing of basic visual information. The hierarchical organization of these systems indicates specialized, though highly interrelated, regions, and a biasing of the competition for feature selection can occur on several of these levels due to feedback connections from higher cortical regions.

Spatial Attention Another way to conceptualize the manner in which attention operates is that it is oriented according to spatial coordinates within a visual scene, instead of, or in addition to, its orientation to specific features. *Spatial attention* can be directed to empty space or objects in space, and does not necessarily have to be driven by features. It can also quickly obtain the gist or summary representation of a visual scene, for example, through a fast and effortless encoding of the spatial layout of objects (for reviews, see Carrasco 2011, Cave and Bichot 1999, Chica, Bartolomeo, and Lupiáñez 2013).

One theory of spatial attention is the *spotlight model*, which proposes that attention tends to be focused on a limited region, for example, to highlight an object at a specific location (LaBerge 1983, Posner 1980). Michael Posner, through a series of cuing experiments, advanced the spotlight theory of spatial attention (Posner 1980, Posner, Snyder, and Davidson 1980). This model describes the orienting of attention as analogous to the operation of a spotlight on a stage, where a focused region of light follows the subject meant to be the center of our attention. Although all of the stage is "sensed" as we view this scene, attention is focused on the main subject of the scene, and we tend to overlook the details of objects outside this spotlight because of limited processing channels. This movable spotlight of attention, however, can be cued to be directed to specific regions and may be made as focused or as diffuse as necessary, at the cost of reducing the amount of detail it can obtain when it is made more diffuse. The *zoom lens* analogy is another way to describe how focused and distributed attention may operate in a visual scene (Eriksen and Yeh 1985). When zooming in, you can pick up more details and features, but when zooming out you tend to get a coarser, more general representation of a scene.

One important characteristic of distributed spatial attention is its sensitivity to the overall properties of a scene, indicating an interaction with feature-based attention. This includes statistical summary representations of scene features, such as the average orientation of tilted line segments, the average size of a set of objects, or the center of mass of such sets. Indeed, recent studies show that a distributed attention can effortlessly obtain a summary representation of the information outside of the primary focus of attention (e.g., Alvarez and Oliva 2008, 2009, Chong and Treisman 2005a, 2005b). For example, even when attention to spatial information is reduced under a taxing dual-task experiment in the laboratory, where the primary focus of the task is to follow moving objects while counting the number of times they have crossed a line drawn across the display, observers can still detect a change in the average orientation of the line segments in the background while continuing to track successfully (Alvarez and Oliva 2009). Such results support the idea that spatially distributed information can be pooled into an 'ensemble' summary representation and remain detectable outside the primary focus of attention, to help overcome the visual processing limits imposed by attention-demanding tasks.[3]

The ability to compute summary representations efficiently is a crucial function of visual processing that seems to operate continuously. Studies suggest that this aspect of spatial attention is an independent form of attention that is always "online," providing this information regardless of whether or not the information reaches conscious awareness. This conceptualization of attention indicates that there must be different forms of spatial attention, some of which, such as these computations of scene statistics, do not necessarily reach conscious awareness but yet can still influence behavior.[4]

Distributed spatial attention tends to process visual information at a lower resolution, or level of detail, and this affects the amount of object-related information that can be encoded into working memory (He, Cavanagh, and Intriligator 1996, Intriligator and Cavanagh 2001). There are studies that indicate a limit to the number of locations that can be selected at once, for example, with a limit of two or three locations when a precise localization is required, and a greater limit of six or seven items when a coarser representation is sufficient (Franconeri, Alvarez, and Enns 2007). Recent studies have also found that a greater number of objects can be localized when the stimulus is presented briefly (e.g., at 50 milliseconds), but at the expense of reduced spatial accuracy in sets with more objects (Haladjian and Pylyshyn 2011, Haladjian et al. 2010). For encoding features in addition to location, such as color or shape, even fewer objects can be processed under brief exposures (Feldman 2007), which indicates how assembling more detailed representations requires more focused attention.

The spotlight and zoom lens models of spatial attention describe the phenomenology of attention and are intuitively appealing, since a single mechanism of attention can be adjusted according to task demands. Yet they do not completely explain other aspects of attention, such as attentional capture. *Attentional capture* is the automatic detection of a stimulus, often task-irrelevant, which occurs when a highly salient stimulus, like a brightly lit object, appears in the visual field (Theeuwes 1994b, 2004, Yantis and Jonides 1984, 1996). Such transitory attractors can be noticed even when outside of the focal spotlight. This suggests that there is always a diffusion of spatial attention in some sense (e.g., to serve feature-based attention related to attentional capture), so it cannot be characterized only as a single spotlight or zoom lens. In other words, focused and diffused attention are not mutually exclusive alternatives; spatial attention has both focal and peripheral aspects that result in differing levels of acuity.

Nevertheless, there does seem to be a continuum to the focus of attention such that it can be adjusted from narrow to broad. Many studies, for example, have shown a systematic interaction between attentional load—the number of concurrent objects or tasks demanding attentional resources—and task performance, such that the more heavily attention is loaded, the less visual information is encoded accurately into memory for later recall (Chesney and Haladjian 2011, Cowan, Blume, and Saults 2012, Fougnie and Marois 2006, Franconeri, Alvarez, and Cavanagh 2013, Fukuda, Awh, and Vogel 2010, Luck and Vogel 1997, Treisman 2006). These studies indicate a trade-off between how much attention is available for encoding information and how well object properties, including locations, can be remembered.

An important form of spatial attention is *covert attention* (for a review, see Carrasco 2011).[5] Covert attention is the voluntary shift of attention outside the center of one's gaze without changing the direction of gaze, typically tested through the spatial cuing paradigm (Posner 1980, Posner, Cohen, and Rafal 1982). That is, the focus of attention can be oriented willfully, and independently from the physical manipulation of gaze or other physical movements. This is in contrast to *overt attention*, which refers to what is typically at the center of your gaze and falls on the fovea of the eye. The evidence of covert attention illustrates how aspects of distributed attention can be highlighted outside the central region of attention.

Covert attention has been shown through various tasks where a subject views the center of a stimulus display but shifts the focus of attention to a target in the periphery without moving her eyes or making other physical movements (Hoffman and Subramaniam 1995, Peterson, Kramer, and Irwin 2004). You can simulate the experience of covert attention by focusing on only one word on this page. While keeping your gaze fixated on that word, try to attend to the location of the page number. Depending on the word's location in relation to the page number, you may be able to read the number, but if the word is too far from it, you will only detect that there is something there even though you are unable to read the number due to the poor attentional resolution available in the visual periphery. Directing your attention to the page number while remaining fixated on the word you chose is an example of covert attention.

This form of attention appears to be particularly important for the planning of eye movements, for example, by covertly detecting where eye gaze should be directed in order to obtain more detailed information from that region (Yeshurun and Carrasco 1998). This attentional shift occurs faster than making an eye movement and tends to favor shifting the gaze toward objects (Horowitz et al. 2004). Covert attention has been observed even in young children, along with developmental improvements in the strategies for directing attention covertly (Enns and Brodeur 1989). This suggests that covert attention is flexible, can be developed with experience, and may be applied to other cognitive processes; this form of attention may be especially important for abilities unique to humans. For example, covert attention may be related to the ability to attend to certain thoughts from memory or to other mental states not immediately linked to sensory information (an idea related to Aristotle's *koine aisthesis*, or 'the common sense,' which describes such an ability to reflect on internal nonperceptual states[6]).

The results from studies in neuroscience confirm that the systems that maintain sustained focused attention and those that shift spatial attention are distinct, but do have some overlapping neural structures. Although the classifications of neural systems supporting various aspects of spatial attention are still under scrutiny, we do have a sense of how the different processes are localized in different brain regions. This neural specialization distinguishes the forms of attention from each other, as we will discuss further in section 2.2. Overall, the research favors arguments claiming that spatial attention is a unique form of attention, dissociated in some of its aspects from either feature-based or object-based attention.

Object-Based Attention Although location-based attention may explain some of the phenomenology of visual experience, various studies characterize objects as the elementary units of attentive processing (Cheries et al. 2006, Feldman 2003, Lee and Chun 2001, Luck and Vogel 1997, Treisman 1988). Under this model, *object-based attention* is drawn to things in the world that display object-like properties, such as cohesion, common fate, and symmetry (for reviews, see Chen 2012, Scholl 2001). That is, the visual system tends to parse the world into objects based on these Gestalt properties and then analyze the objects with more detail via focal attention (Neisser 1967).

Selective attention appears to operate on objects as whole entities and in parallel, including when these objects are competing for selection (Desimone and Duncan 1995). The richer aspect of object-based attention typically is thought to be a stage of selective attention that occurs after a parallel preattentive process that segments a visual scene into discrete objects (Duncan 1984). Support for object-based attention is shown by an observed 'within-object advantage' of feature detection: target features that appear within a previously cued object are more efficiently processed, compared to the processing of the same feature should it appear on another object or in empty space at the same distance (Egly, Driver, and Rafal 1994). This suggests that attention is automatically spread within an object when a cue draws attention to that object. (See figure 2.3 for examples.) The within-object advantage is also observed when subjects are asked to report on object representations in working memory, that is, after the stimulus has disappeared (Awh et al. 2001). This tendency for attention to operate on objects probably exists because, with some exceptions like substances such as water or sand, our perceptual world is mostly comprised of discrete objects.

The number of objects that can be selected in parallel tends to vary, depending on the design of the experiments, but recent evidence suggests that up to 12 objects can be segmented in parallel as "potential units of selection" for further attentive processing (de-Wit et al. 2011). The level of detail encoded for each object, however, is limited by other cognitive mechanisms such as working memory capacity (Franconeri, Alvarez, and Cavanagh 2013). Therefore, most theories on object-based attention identify the

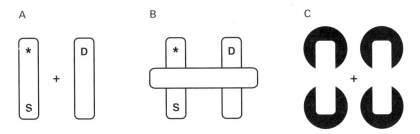

Figure 2.3
Examples of stimuli used to test object-based attention by identifying a within-object advantage. When an empty object is first cued at a specific location, for example, by changing the color of the outlines on one end of the rectangle (represented by the asterisks in this figure), observers are subsequently faster when detecting a target feature (such as identifying a change in luminance or a letter) that appears on the same object (at location 'S') than one that appears at the same distance but on a different object (at location 'D'). This within-object advantage occurs under various conditions, including when part of the object is occluded (middle example) or when the object consists of illusory contours and requires perceptual completion (example on the right). This image is based on figure 3 in Scholl (2001), where several experiments testing the within-object advantage are described (e.g., Egly, Driver, and Rafal 1994, Moore, Yantis, and Vaughan 1998).

capacity limits of these "simultaneous" representations based on various factors, including how much detail must be encoded and the context.

At least two perceptual functions are crucial for object-based attention: individuation and identification. The first of these, individuation, is a data-driven process that selects object-like items for further processing and is thought to occur within a cognitively impenetrable module referred to as *early vision* (Pylyshyn 1999). The second identification stage requires the binding of features from attended objects into rich mental representations and allows object identification or recognition (Ballard et al. 1997, Kahneman, Treisman, and Gibbs 1992, Treisman 1996, 1998). Together, individuation and identification contribute to the experience of attending to and recognizing specific things in the world, and supports abilities such as enumerating sets of items, tracking multiple objects, and focusing on a single item in detail.

Object individuation relies on the coherence of visual input that extends over space and time (Feldman and Tremoulet 2006, Scholl, Pylyshyn, and Feldman 2001). Coherence is influenced by the spatial relationships among visual stimuli, including geometric factors such as symmetry, good continuation, and parallelism (Feldman 2007). Segmentation processes in early vision operate to divide visual arrays into distinct objects in accord with the Gestalt principles of cohesion, boundedness, rigidity, common fate, and "no action at a distance"—meaning that distant objects appear to be independent from one another (Spelke 1990). Preattentive processes, such as 'visual routines,' assist individuation by extracting spatial relations and other properties from the output of early visual processing, which also helps in tracking objects or marking locations (Ullman 1984). By dividing visual scenes into distinct objects, preattentive processes form an organizing structure that allows attention to bind features in visual short-term memory (Feldman 2003, Wheeler and Treisman 2002). This sort of bottom-up processing must operate on a scene before features can be encoded or consciously perceived.

Zenon Pylyshyn (1989, 2001, 2003b) describes a possible mechanism responsible for individuating visual objects in his *visual indexing theory*. On this view, 'visual indexes' are pointers to objects in a visual scene that are maintained over temporal and spatial changes (also named 'FINSTs' for 'Fingers of INSTantiation').[7] These data-driven and preattentive pointers are triggered by the object-defining properties that parse a discrete object from a visual scene. Once a pointer is assigned, it can "stick to" and follow a visual object (or proto-object, at this stage) without necessarily encoding any of its properties. This individuation of visual objects is related to *demonstrative thoughts* (i.e., the ability to think about and refer to *this* or *that* thing) and provides a solution to the 'reference problem' that is concerned with how it is possible to track an object through space and time while maintaining the corresponding link to its mental representation (Perry 1997, Pylyshyn 2003b, 2007, Siegel 2002). In this way, early attentive processes help anchor mental representations in the world. These visual indexes also serve localization and enumeration abilities (Chesney and Haladjian 2011, Haladjian and Pylyshyn 2011, Trick and Pylyshyn 1993, 1994), facilitate visual search (Burkell and Pylyshyn 1997), and are considered the first step in solving the binding problem by providing the core structure for building rich object-based representations (Treisman 1996, Treisman and Zhang 2006, Wolfe 2012, Wolfe and Cave 1999).

A paradigm for testing the visual indexing mechanism is the *multiple object tracking* (MOT) task, where observers track several target objects that move unpredictably and independently among identical distractors (Pylyshyn and Storm 1988). Research suggests that we can typically track between two and five of these moving objects successfully for extended durations and under varying conditions, including when targets move behind occluders or change in shape and size. Although multiple objects can be tracked successfully, their identities are often confused, since this primitive low-level mechanism does not encode details of the objects being tracked (Pylyshyn 2004). The task of feature binding is reserved for higher-level processes.

After objects are indexed, a serial attention can then operate on the indexes in order to form more detailed representations in visual short-term memory, or to switch between objects during situations where moving targets may be confused with nearby distractors. That is, serial attention uses these pointers as guides for where to focus to obtain more visual information, as is shown by an increased sensitivity to feature changes on tracked objects (Sears and Pylyshyn 2000). Attention can be shifted quickly among visual indexes depending upon task demands (Pylyshyn 1989), and this attention can help build richer representations by integrating features into object representations. This does not necessarily indicate that a "divided" attention is at play during this stage, but rather that an ability to move serial attention quickly among indexed objects is involved in the process.[8] Consequently, the limit of how many items can be tracked depends upon attention-demanding scene characteristics such as the speed of the objects and the amount of spatial interference among the moving objects (Holcombe and Chen 2012), which will tax the ability to deploy serial attention or maintain indexes.

This brings us to the second stage of object perception, the *object identification* stage. *Object file theory* (related to feature integration theory) describes how object-based attention may operate to bind features via midlevel 'object file' representations that aid the identification process (Kahneman, Treisman, and Gibbs 1992, Treisman and Gelade 1980). Selective attention plays a crucial role in the formation of persisting object representations by allowing scene features to build a coherent representation incrementally (Treisman 1998, 2006). This binding happens quickly; as little as 200 milliseconds can be sufficient to encode much information (Feldman 2007). The efficiency of this process, however, is sensitive to cognitive load—it becomes more difficult when attention or working memory resources are preoccupied with other tasks or multiple object representations (Johnson, Hollingworth, and Luck 2008, Wheeler and Treisman 2002). Several infant studies provide support for early instances of the integration of features into meaningful representations, but only after a certain developmental stage (Xu and Carey 1996). These object files have also been used to explain the developmental ability to count sets of objects (e.g., Carey 2001, Feigenson and Carey 2003). Such studies suggest that some forms of attention can become more sophisticated with experience.

The construction of object representations likely occurs in visual short-term memory, which is limited in the number of objects and the amount of featural information it can represent (Alvarez and Cavanagh 2004, Xu and Chun 2006). We seem to have a capacity of around four object representations, a common limit identified in research on working memory (e.g., Cowan 2001), although the idea of a fixed limit has been challenged recently because it tends to vary depending on the experimental setup (see Alvarez and Cavanagh 2004, Franconeri, Alvarez, and Cavanagh 2013).

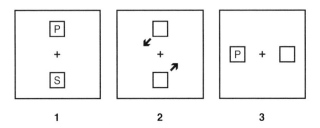

Figure 2.4
Object file theory was supported by experimental results that showed a preattentive phenomenon called the *object-specific preview benefit*. Here, observers responded faster to a target stimulus when it reappeared in the same object that had briefly previewed it, even after the stimulus moved through space (compared to cases when the target appeared at the same distance but in empty space, or appeared in a different object that previewed a different letter). This result emphasized the object-specific nature of the preview effect, as opposed to being location specific. The figure above illustrates the original preview experiment with motion (adapted from Kahneman, Treisman, and Gibbs 1992). Panel 1 shows the initial display with preview letters that appeared briefly. Panel 2 depicts the movement of the empty objects that lasted 130–590 milliseconds. Panel 3 shows the test letter (target) that was to be named. In this example, a 'congruent-matching' trial is depicted, which would produce the fastest reaction times since the same letter (matching) appeared in the original preview object (congruent).

This two-stage process, by which a selective attention can bind and maintain features in an object file but only after the individuation stage, exemplifies the interaction between low-level and high-level forms of attention that makes the study of attention so complex. See figure 2.4 for more on the object file theory and a description of the experiment that identified it.

Recently there has been a move toward characterizing the limited-resource problem, referred to in the literature as the 'information bottleneck,' as being determined by a map-based continuous and *flexible resource* (e.g., Brady and Tenenbaum 2013, Franconeri, Alvarez, and Cavanagh 2013). This perspective is in contrast to more traditional discrete *slot-based models* which posit that information is encoded into a fixed number of available slots in working memory (e.g., see Cowan 2001). The hierarchical nature of the spatially organized feature maps in the visual cortex presents different levels of resource limitation, depending upon factors such as spatial resolution and the number of features that are encoded and required for object recognition (Alvarez and Cavanagh 2004). This flexible map-based resource approach is not as rigid as traditional slot-based models, and it seems to better reflect the observed variations in the allocation of resources during attentive tasks, including the individuation of objects (i.e., it seems that both target and nontarget objects are individuated).

Various studies support the disassociation between individuation and identification stages of attention. For example, a study on 12-month-old infants found that the feature information used to individuate objects is not always included in its representation (Tremoulet, Leslie, and Hall 2000). The finding that features can help individuate objects but need not be encoded into a persisting representation of those objects is consistent with visual indexing theory: the indexing mechanism operates prior to the deployment of serial attention, therefore it can be active prior to the encoding of the features from an indexed item. It identifies only that something is there, not what it is.

The object-based attention model proposes that the visual system evolved to favor the selection of whole objects in a visual scene as opposed to properties such as location or individual features, and many studies provide support for an object-based theory of attention (Baylis and Driver 1993, Chen 2012, Kahneman, Treisman, and Gibbs 1992, O'Craven, Downing, and Kanwisher 1999, Scholl 2001, Scholl, Pylyshyn, and Feldman 2001, Shomstein and Behrmann 2006, Treisman and Zhang 2006, Wolfe and Bennett 1997,

Yantis 1992). In addition to the selection of relevant objects, the *inhibition* of irrelevant items in the visual scene has been shown to be object based (Geng 2014, Pylyshyn 2006, Theeuwes and Godljn 2002, Theeuwes, Van der Stigchel, and Olivers 2006). In fact, such inhibition, which prevents the further processing of information, is specific to the irrelevant objects selectively marked for inhibition and does not occur in the empty space near those objects (Pylyshyn 2006, Pylyshyn et al. 2008), nor does it affect nontarget objects clearly segregated within three-dimensional space that are not likely to cause tracking interference (Haladjian, Montemayor, and Pylyshyn 2008). These selection and inhibition processes allow attentional capture to occur in the empty space between moving objects but not on the nontarget objects that are being inhibited, since those distractors may interfere with the tracking of the target objects and thus require inhibition. This evidence supports the notion that low-level object individuation tends to operate on both target and nontarget objects. Such studies emphasize the importance of object-based perception, which is likely an important adaptation in visual information processing systems to enable complex interactions with discrete objects.

One perspective on object-based attention, which suggests that it could be a more recent adaptation, argues that this form of attention is observed more often under conditions of high uncertainty (Shomstein 2012). When there is high certainty about scene characteristics—for example, when the spatial location of a target is known—then a focused spatial attention is prioritized instead of an object-based strategy, since the latter tends to select multiple objects in a scene that may or may not be targets. This 'attentional prioritization' is a flexible resource, applied strategically on the basis of task demands and perceptual conditions. This model follows an attentional load theory, where more uncertainty requires more attention, and explains performance on a graded continuum.

Such diverse theories and studies give us good reason to believe that different attention systems work in complementary ways to process perceptual information, and that object-based attention is an evolutionarily newer processing strategy that developed after the more basic feature-based and spatial forms of attention. The usefulness of the object-based model of higher-level representation is that it provides a structure wherein low-level information from the various visual pathways can be integrated to form a coherent and persisting representation of a visual object (Ballard et al. 1997, Kahneman, Treisman, and Gibbs 1992, Noles, Scholl, and Mitroff 2005). Some studies, however, suggest that this binding can happen even when pairs of features are simply superimposed spatially (Holcombe and Cavanagh 2001) and thus not necessarily bound in an object file format. Nevertheless, such forms of 'conjunction attention' enable the crucial integration of multiple features. This ability is particularly important for guiding actions and for conscious attention, which we discuss further in chapter 4. Our position is that before you can have a conscious representation, visual information must be organized in some useful way. Object file representations provide this organization, especially since visual features usually belong to discrete objects. Without the ability to select an individual object and bind its features, an agent could not sustain a persisting representation of the object.

We are particularly drawn to the object-based attention model because it provides a nice structure for the integration of information from the various visual subsystems to form a coherent representation of a visual scene, in a way that allows mental representations to refer to external objects. Whether or not mental representations truly are organized via object files remains debatable, but for our purposes the form in which features are integrated is not problematic as long as there is some account for this integration. We believe that object files are theoretically important for providing the content of mental representations and for integrating perceptual information from multiple modalities, as well as from other forms of attention, in order to produce a coherent representation. It is these representations that most likely make up the contents of conscious experience.

2.1.2 Bottom-Up versus Top-Down Attention

The varieties of attention that we described functionally in the previous section can be further categorized by distinguishing two ways in which attention can be controlled: in a *bottom-up* manner or a *top-down* manner (for reviews, see Corbetta and Shulman 2002, Mulckhuyse and Theeuwes 2010, Theeuwes 2010, Wright and Ward 2008). That is, some aspects of attention can be thought of as bottom-up, feedforward neural processes that are stimulus driven, automatic, exogenous (triggered by external stimuli), and cognitively impenetrable—that is, not susceptible to willful control. On the other hand, some forms of attention are more goal-oriented and can be controlled by top-down processes that originate endogenously (e.g., controlled from higher cortical areas), influenced by feedback or recurrent neural systems, and are more effortful and deliberate. This form of 'active attention' is the more obvious form of attention, a quality that motivated its study early on (e.g., see Bradley 1902, James 1905/1890).

The investigation of bottom-up attention, which expanded in the 1950s, was influenced by *information processing* theories that had emerged in computer science and communication systems research (Broadbent 1958, 1977, Bundesen 2001, Miller 1953, 1956, Pylyshyn 1965, 1984, Shannon and Weaver 1949, Sperling 1960, Theeuwes 1993). This computational perspective—based on the idea that the brain works like a computer—continues to influence the field and emphasizes the capacity limits that face attention at various stages in the processing of visual information, characterized by observed 'bottlenecks' in the information that can reach awareness. For example, when a scene includes multiple objects that must be attended to, as when you are navigating your bicycle through the busy streets of Amsterdam or trying to find that red pen on your cluttered desk, the visual information from these detail-rich scenes must compete for your attention before it can reach awareness. Therefore, the purpose of attention is to serve as a filter that selects the most relevant information. But how is information selected, and at what stage? In low-level input channels, or in higher-level temporary memory stores? To address those questions, some principled theories of selection have been proposed.

Donald Broadbent's *filter theory of attention* was concerned with the processing of auditory information, and he is considered one of the first to apply the concepts of information theory from computer science to cognitive psychology (Broadbent 1952, 1958, Broadbent and Gregory 1963). A goal of his work was to model the ability to attend to a specific sequence of sounds amid a variety of auditory inputs that together sound like a cacophony of noise. This ability is at play in the *cocktail party effect* identified by Colin Cherry (1953): the phenomenon of being able to selectively hear our own name in the otherwise noisy stream of auditory inputs typical at a crowded and lively cocktail party. The selective attention model proposed by Broadbent (1958) is an *early selection* model, since it relies on the early stages of visual processing where only the physical properties of the stimuli are present, as opposed to a higher-level stage when one has semantic information about the stimulus or is able to recognize it. The main claim here is that a large amount of sensory information is transmitted through the early

Forms of Attention

preattentive input channels in parallel (simultaneously), but this uncategorized information is reduced or limited before it reaches a second stage, as it is buffered and selectively filtered for encoding into capacity-limited short-term memory storage. This selection was thought to occur in an 'all-or-none' manner, where only some information passing a certain threshold enters the second processing stage.

George Sperling, testing the observation that "more is seen than can be remembered," showed that more visual information is processed by early vision than reaches conscious awareness (Sperling 1960, 1963, 1989).[9] In a series of experiments testing iconic memory for briefly presented arrays of letters (in this case, three rows with four random letters in each row), Sperling found that observers could respond with higher accuracy when asked to report only a partial subset from the array of letters (for example, the third row of four letters) than when asked to free report from all three rows of letters (Sperling 1960). The observers were cued after the stimulus presentation as to which of three rows to report upon, and still were able to respond with high accuracy. This result suggests that the information from all three rows was available in iconic memory; but evidently, this information can fade quickly, since being asked to report all the letters led to more errors.

Such studies from two-stage models support the idea that the first stage tends to have a large capacity that operates in parallel. The limitation on the amount of information that can be processed is determined by the next, 'late selection' level, which is more selective and requires more resources, thus limiting capacity. As hypothesized in Broadbent's (1958) *all-or-none theory*, the attention system selects a target that is searched for by the allocation of more focused resources toward visual stimuli that are already cued in parallel. Without this prior parallel selection to provide material for sifting, a shift of attention during a target search cannot occur (Jonides 1983).

In contrast, Treisman (1960, 1964, 1969) argued that attention does not operate in such an all-or-none fashion. Instead, items that were more *probable* as targets are filtered at an earlier stage. Framed within Broadbent's filter theory, this would suggest that there is a bias for recognizing stimuli that are more likely to be relevant, depending on the context. Treisman's *attenuating model* served as a contrast to the all-or-none filter theory by focusing on a graded approach to attentional processing, introducing the notion of probability into perceptual systems. Simply stated, the selection mechanism attenuates different stimuli at various processing levels, rather than simply blocking them from advancing past the initial stages of visual processing (Treisman 1960). This allows for the possibility that more stimuli are processed than one is conscious of attending to, and would account for the cocktail party effect: you are generally not attending to the stream of background noise until you notice your name, but nonetheless, the background noise is being processed at some level before your name is recognized. For a stimulus to reach consciousness, a certain level of activation must be achieved, and that activation likely is attenuated by the signals from various feedback networks that help guide attention.

Researchers have studied the problem of information selection, that is, the conditions under which a target stimulus can be detected amid distractors (or 'noise'), by applying *signal detection theory*. This theory too was related to information theory developed for communication and radar devices, but has been applied to perception as another way to formalize information processing (Krantz 1969, Swets, Tanner, and Birdsall 1961, Verghese 2001). Signal-detection theory may be important for determining what information reaches awareness, as it can model how perceptual judgments are made by quantifying the amount of information that is processed. This is also useful for determining the threshold required for information to be detected.[10] Notice, however, that information processing is a more general cognitive event than the access to content such as semantically characterized features of objects. We discuss this issue in more detail in chapters 3 and 4.

Based on these early theoretical models of visual information processing, bottom-up attention can be ascribed to systems in early vision, that is, the low-level systems that select salient visual information automatically. These are core processes in visual attention at a stage that is considered preattentive and cognitively impenetrable (but for arguments challenging the idea of cognitive impenetrability, see Vetter and Newen 2014). Bottom-up attention includes feature-based attention and the forms of spatial attention we described earlier in this chapter. These processes generally initialize perception by serving as input channels for sensory information, and they provide the foundation for detecting and recognizing visual objects. In other words, the visual scene must be parsed through bottom-up mechanisms before properties can be encoded into meaningful representations (Rensink 2000).

For example, the first steps required for object representations are the establishment of a reference to an external object and the detection of the object's features by the specialized subsystems in the visual cortex. These steps encounter information processing limits, since not all information present in a visual scene can be detected and represented at once. According to a model of Koch and Ullman (1985), scene parsing follows a "winner-takes-all" neural process, where the most active neural units in the early visual system bring certain feature clusters, like luminance or motion, into a *saliency map*. Feature-saliency maps are context dependent, and saliency judgments rely on the relative contrast within a specific scene (Itti and Koch 2001). From these feature maps, information can be organized to form a detailed object representation. The importance of these theories is that they attempt to describe how more complex information processing systems allow perceptual information to reach higher levels of cognitive processing.

One way that selective bottom-up information processing can be described biologically is through the *biased competition model* of attention, which proposes that the selection of visual information for attention is the result of neural competition between different items in the visual field (Bundesen 1990, Desimone and Duncan 1995, Kastner and Ungerleider 2001, Koch and Ullman 1985). Typically, the strongest neural response is computed locally and compared relative to other neural activations (Knudsen 2007, Koch and Ullman 1985). Note that context—the surrounding perceptual information—again is important for these more primitive attentional mechanisms. The items that win this competition by exceeding a certain threshold then move forward through the higher pathways of cognition. For example, bottom-up attentional processes can allow the most visually salient properties, such as high luminance, to reach higher-level processing by capturing attention (Donk and Theeuwes 2003, Itti and Koch 2000, 2001, Theeuwes 1994a, 2004), a phenomenon related to the pop-out effect.

As we mentioned earlier, attentional capture describes the automatic detection of visual information through the unselective capture of focal attention. This capture is based on salient features that exceed a threshold of bottom-up 'difference signals' computed from local perceptual information (Burnham 2007, Theeuwes 1992, Yantis 1993, Yantis and Hillstrom 1994), and generally happens at the preattentive parallel stage of processing (Lu and Han 2009, Mathôt, Hickey, and Theeuwes 2010). That is, attentional capture is an automatic shift of spatial attention that relies on saliency computations in the early stages of vision, without any intervention of willful top-down modulation. Whatever is most salient in these initial processing stages of attention can capture attention, and this can occur without awareness (Zhaoping 2008), while also influencing motor commands such as eye movement (Zhaoping 2012). Furthermore, there is evidence that this form of exogenous attention can be object based and not simply tied to the locations on the retina that were stimulated by the salient feature (Theeuwes, Mathôt, and Grainger 2013). Other characteristics of stimuli can capture attention; something that is moving toward you, for instance, is more likely to capture your attention than something moving away (Franconeri and Simons 2003). The general conditions under which a new object will capture attention, however, seem to require a threshold of changes in luminance in order to be detected (Franconeri, Hollingworth, and Simons 2005).

In terms of how memory can affect bottom-up selection, Shimon Ullman (1984) proposed the idea of *visual routines*, which are certain operations stored in memory that can execute scene parsing by extracting spatial relations and various properties from the raw input of early visual processing in a task-dependent manner. Such visual routines assist in tracking contours, counting objects, or marking locations. Some visual routines, called 'sprites,' can also detect certain forms of typical motion patterns, in order to facilitate interaction with the objects that tend to have those motions, such as a closing door or a falling object you must catch (Cavanagh, Labianca, and Thornton 2001). These stored visual routines assist interactions with common moving objects, but a focused spatial attention is required to discriminate these familiar motion patterns.

Overall, bottom-up processes are essential for moving visual information into higher-level cognition, which can provide the contents for mental representations. Even though features that are processed preattentively do not always reach awareness, they still can influence motor commands and conscious report (for a review, see Mulckhuyse and Theeuwes 2010). In addition to providing the starting point for perception, these bottom-up mechanisms address the philosophical question of how the mind can refer to external objects. That is, how does the mind maintain a reference to a visual object such that its features can be encoded correctly, maintained over space and time, and eventually recognized and acted upon? The role of indexicals or demonstratives has been suggested by philosophers and cognitive scientists to address this reference problem, and some solutions, as we mentioned earlier, have been proposed related to object-based attention.

The contrast to bottom-up attention is *top-down attention*. Here we have a more goal-directed deployment of attention, be it to specific features, regions in space, or objects. This sort of attention is more deliberate, often creating a feeling of willful engagement, such as the directing of the attentional spotlight around a visual scene. This form of selective attention is thought to be a stage of focal attention that follows the parallel preattentive process that segments a visual scene into discrete objects (Duncan 1984, Neisser 1967).

For object representations to persist, a selective attention must be directed to the items so that features can be encoded and maintained in a unified representation (Treisman 2006). Although the bottom-up processes described above are considered low level and often occur automatically, the selection of features can be biased by top-down processes. For example, task demands (such as searching for a specific feature in a scene) can influence the outcome of bottom-up competition by directing focused attention toward specific information (LaBerge et al. 1991). Additionally, top-down processes can override automatic processes through intentional control of the visual system, preventing the deployment of attention to irrelevant items (Jonides and Yantis 1988, Leber and Egeth 2006) by inhibiting those items or their locations (Koch and Ullman 1985, Tipper, Driver, and Weaver 1991), or by modulating attention to certain features (Corbetta et al. 1990). The efficiency of this type of top-down modulation is sensitive to various factors related to the strength of bottom-up signals, such as feature saliency (Kastner and Ungerleider 2000) and priming effects (Theeuwes 2013).

When the target is not easily distinguishable from its surroundings or contains a conjunction of features, then a more deliberate and serial search procedure is required, which involves more cognitive resources (Wolfe and Horowitz 2004). The later stage that deploys serial attention is especially crucial when the target does not easily pop out or has a conjunction of features as, for example, in the test of trying to find a blue vertical rectangle among distractor rectangles of different colors and orientations (Treisman and Gelade 1980). *Visual search*, where the observer's task is to find an item with target features among distractors, is one of the main paradigms for studying this top-down aspect of feature-based attention (Pashler 1998, Peelen and

Kastner 2014, Treisman 1982, Treisman and Gelade 1980, Wolfe 1994, Wolfe, Cave, and Franzel 1989). *Guided search theory* is one model of serial top-down attention, describing a systematic search through visual information influenced by task demands (Wolfe 1994, Wolfe, Cave, and Franzel 1989). The guided search model proposes a two-stage process that involves a parallel processing of different kinds of features, modulated by bottom-up and top-down signals to guide the deployment of attention toward a target feature. This exemplifies the difference between processing visual information in parallel, when simple features are easily processed, or serially, when more attention is required to detect the target, such as when the target is made up of a conjunction of features or surrounded by similar-looking distractors.

John Tsotsos and colleagues (1995) propose a *selective tuning model*, in which attentional processes selectively tune the neural processing network to determine which scene features are visited, in order to optimize the matching process between the target features or representations and the perceptual inputs during visual search. A preattentive parallel system identifies likely candidates, which are selected when a top-down, winner-takes-all process identifies the likely match (Tsotsos et al. 1995). Again, this demonstrates how attention relies on the interaction between bottom-up and top-down flows of neural signals when filtering relevant information from the plethora of perceptual input that faces the visual system.

Some studies suggest that top-down attention may enhance not feature-based attention itself but rather the allocation of spatial attention during visual search (Moore and Egeth 1998). This account of attention prioritizes spatial properties over featural properties (Liu, Stevens, and Carrasco 2007) and highlights the role of location for selection. Such perspectives claim that top-down selection relies on *where* attention is directed in space, as in the spotlight model, and on *how well* spatial filtering can be implemented (Cave and Bichot 1999, Eriksen and Yeh 1985, Nissen 1985, Posner, Snyder, and Davidson 1980). Regardless of whether it operates on features, on objects, or on space, top-down attention tends to modulate the processing of information based on goals or willful selection.

The study of top-down attention will eventually lead us to aspects of visual working memory and short-term memory. There is evidence for interaction between selective attention and working memory in the presence of top-down modulation, involving some shared neural systems (Awh and Jonides 2001, Gazzaley and Nobre 2011). Another proposed mechanism related to working memory is the *central executive* system (Baddeley 2000, Baddeley and Della Sala 1996, Baddeley and Weiskrantz 1993). The central executive is thought to assist in the maintenance of object representations in visual working memory by way of various attention and working memory processes, such as the visuospatial sketchpad, the episodic buffer, and the phonological loop (Baddeley 2000). These memory structures may also support self-reflection and help distinguish the source of phenomenal experience, be it from sensory inputs or from long-term memory (Johnson, Hashtroudi, and Lindsay 1993, Mitchell and Johnson 2009). Together, these components describe ways in which perceptual information is maintained for cognition; that maintenance may be necessary for conscious awareness of the information.

Indeed, it is likely that multiple mechanisms are involved in the ability to attend to representations in working memory. Recent studies indicate that spatial attention is crucial for maintaining features in working memory (Williams et al. 2013). Visual working memory also serves attentional processes by modulating the competition of information selection at different levels (e.g., saliency is lower-level, while shape perception is higher-level) and follows the constraints typical of information processing systems (Knudsen 2007). The relationship between working memory and attention is a crucial one, but we can only mention it briefly here (for overviews, see Gazzaley and Nobre 2011, Hollingworth and Maxcey-Richard 2013, Oberauer and Hein 2012, Olivers 2008).

This distinction of bottom-up versus top-down attention, however, does not describe all the various forms of attention. Recently this dichotomy has been challenged, since there are other influences on attention that do not fall neatly into these categories. Learned rewards or habits, for example, have been shown to influence attention (Awh, Belopolsky, and Theeuwes 2012). Some aspects of cognition (e.g., those based on previous experience) also can bias how attention selects information—even though the information selected as a result of that influence may not be relevant to the task at hand—and such selection may not be a willful, top-down deployment of attention.[11] Although the bottom-up versus top-down classification of attention is not a perfect dichotomy, many processes within the broad category of 'attention' can be described as behaving in one way or the other, and this distinction can help guide our understanding of how attention operates with respect to what reaches consciousness.

In sum, it is generally accepted that both bottom-up and top-down processes modulate attention depending on the specific situation (Connor, Egeth, and Yantis 2004, Treisman 2006, Watson and Kramer 1999). Such interactions produce our experience of focused attention and awareness. It is worth noting here that this subjective experience of 'attending to something' is comprised of the several forms of attention that have been described so far. By now you should be convinced that studying attention is not a straightforward endeavor. Although some argue against a strict dichotomy of bottom-up versus top-down modulation of attention, the delineation does serve the purpose of differentiating those processes that are lower-level, and perhaps more primitive in the evolutionary sense, from those that are part of higher-level cognitive control and more evolutionarily advanced. It is especially important to examine the systems involved in top-down attention, since they may play critical roles in the integrative processes necessary for conscious attention, for example by focusing attention to objects in order to better encode features. We will talk more about these systems in section 2.4.

2.1.3 Effortless versus Effortful Attention

The distinction between effortless and effortful forms of attention will mirror much of what we just discussed, and provide further nuances to consider regarding the relationship between attention and consciousness. *Effortless attention*, like bottom-up attention, is described as an involuntary form of attention, and much of it does not reach conscious awareness. On the other end of this spectrum is *effortful attention*, which, like top-down attention, is described as focused, deliberate, voluntary, or goal-driven and produces the subjective feeling of expending effort.

This distinction, however, like most proposed dichotomies, is not so straightforward. Some complex attentional processes can be so engrossing that they produce the subjective feeling of being involved in a task so effortlessly that one loses a sense of time (see Bruya 2010). It is this latter version of effortless attention that is particularly interesting; it is related to expertise and is suggestive of how memory systems can interact with attention to influence the perception of effort and time. Such phenomena also illustrate the complex interactions among attentional systems as we attempt to categorize them as distinct processes.

Brian Bruya describes attention as a "mechanism of sensitization that draws information relevant to dynamic contextual structures of reference through dynamic processing pathways" (Bruya 2010, 11). A focused attention may operate, in this view, by inhibiting "intrusions" from competing sources of dynamic information (Bruya 2010, 12). Effortless attention occurs when these competitions stabilize, resulting in a predominant source of information, which then does not have to compete with the others. That is, under such "sensitized" conditions there is no more need to actively inhibit irrelevant information, and this leads to a more efficient and effortless processing. The question then arises: what does the inhibiting? Inhibitory mechanisms are proposed in the attention models we have discussed so far—such as those within object-based attention and winner-takes-all selection processes—but do those mechanisms function the same way to support effortless attention as proposed by Bruya? Inhibition is important to consider for understanding attention, since attention not only selects certain information to reach higher levels in cognition but also serves to inhibit other information from doing so. One purpose of such inhibitory attentional processes may be to prevent certain information from entering conscious awareness; those processes exemplify forms of attention that remain outside of awareness yet still can directly affect it.

Bruya argues that effortless attention is crucial in some aspects of conscious awareness, such as the experience of free will or flow.[12] It is also thought that this form of attention may be more closely tied to working memory or to an executive central processor (Bruya 2010, 122–3). Yet another view says that some forms of effortless attention are derivative from the motor control systems (Goodale 2011), suggesting the involvement of learned associations; but this is difficult to reconcile with the robust finding that all motor control is unconscious and thus not experienced as a form of effortless attention.

Concerning the influence of motor control, Bernhard Hommel (2010) proposes that attention is a "by-product" of the distributed processing systems that control action, and that perceptual systems provide the parameters required for the action systems. This places perceiving and acting close together, which is important to point out since attention enables many kinds of behaviors, whether motor actions or thoughts. On this view, attention is typically effortless unless there is something that conflicts with the 'default' action system and requires additional cognitive processing (Bruya 2010, 11). In any case, it becomes apparent that the distinction between effortless and effortful attention is not a clear low-level versus high-level dichotomy: effortlessness can be related to the brain's default state, and many high-level processes can become effortless. The implications of this insight for consciousness are important to consider, and we will have more to say about them later.

The main idea to take from this effort-based conceptualization of attention is that attention operates in a complex and intricate way. Even a form of attention considered to be effortful can *become* effortless, as the subjective nature of the activity, which results from these processing systems, can change with experience (e.g., expertise), and attention can be implemented in different ways depending upon the available information or conditions. Regardless, the many mechanisms of attention described in the previous sections all play roles in both effortless and effortful attention, but they do not always fall into one or the other category. Like bottom-up versus top-down, the effortful and effortless forms of attention do not form a clear-cut classification.

The varieties of attention we have discussed so far ultimately attempt to describe visual information processing and selection, but not necessarily conscious experience. The distinct functions of these different processes—feature detection, object tracking, object identification—should be supported, in theory, by distinct neural mechanisms. Of course, the literature suggests a more complicated relationship between these forms of attention, with top-down and bottom-up systems interacting closely with other resources, such as memory, to achieve the goals of the organism. Nevertheless, examining the neural substrates of attentional processes helps us clarify which systems are more basic and which are more complex. That clarity is especially useful for our understanding of attention from an evolutionary perspective, and for determining its relationship to consciousness.

2.2 Evolution of Attention (as Functionalism) and Neural Structures

The functions of attention can be described in adaptive terms, although little theoretical and empirical work has been devoted directly to an evolutionary account of attention (see Cosmides and Tooby 2013, Tooby and Cosmides 1995, Ward 2013, Wright and Ward 2008, 235–41). For example, the most basic form of attention—the one that is more stimulus-dependent or bottom-up—is shared by most species, including insects (e.g., see Wiederman and O'Carroll 2013, Wiederman, Shoemaker, and O'Carroll 2008). Many animals also are capable of a selective attention for feature and object recognition within a sensory modality similar to the human visual sense. From an evolutionary perspective, being able to attend to various aspects of the environment is so critical for survival that it must be physically instantiated by various different neural systems across species. The basic cognitive functions related to motor control and navigation, for example, require attention to select relevant features of the environment. Even "simpler" organisms like insects can locate the position and angle of the sun and determine orientation or perform computations on how the location of external landmarks relates to their egocentric frame (Gallistel 1998, Wang and Spelke 2002). Bees in particular have a sophisticated navigation system for finding sources of food and for communicating this information (through their 'waggle dance') to other members of their hive. These functions are clearly adaptive and serve the purpose of increasing an organism's fitness for survival.

The same brain areas that perform these more primitive attentional sorts of behaviors are consistent and present among various species (e.g., reptiles, fish, amphibians, mammals), with especially relevant roots in systems present in vertebrates since the Cambrian period (Feinberg and Mallatt 2013). One advantage of studying the brain from an approach which assumes that natural selection tends to produce systems that solve adaptive problems is that it allows researchers to develop testable hypotheses about information processing systems in the brain that make up 'attention' (Cosmides and Tooby 2013, 204). These considerations regarding the evolution of attention may also indicate how conscious awareness evolved in humans (Merker 2005, 2007a, 2007b). For example, the higher-level brain regions that appear to correspond to conscious awareness are associated with integrating the sensory inputs from the various modular systems in the visual cortex. By further examining the neural structures that support attention from this perspective, we can get a better idea of how attention and consciousness may be related.

It is uncontroversial that different areas of the brain developed at different times throughout evolution. As we will further discuss in section 5.1, there is good reason to believe that attention is an early adaptation in the evolution of the human mind (Tooby and Cosmides 1995). For example, "older" sensory-motor brain areas that evolved early in vertebrates include the cerebellum and hippocampus, which are crucial for motor control, navigation, and vital body functions in humans. The limbic system, better developed in mammals but also present in birds and some reptiles, includes the amygdala, hippocampus, and hypothalamus. These are systems that aid in memory formations (i.e., encoding into long-term memory), produce fear and reward responses, and support spatial memory. The lateral geniculate nucleus (LGN) moves information from the optic nerve into the visual cortex (including areas V1–V5), which is the earliest part of the cortical brain. The cerebral cortex is the more "recent" part of the brain, the frontal cortex being the newest, and it enables more complex behaviors such as action planning, problem solving, and language. Structures related to the cortex can be traced back to the forebrain of vertebrates and even to neural structures within invertebrates (Tomer et al. 2010). Nevertheless, one can argue that the increasing complexity of organisms is correlated with an increased complexity in the organization (and size) of the brain, particularly those areas related to attentional systems; this correlation has evolutionary implications regarding cognitive abilities such as language (Bor and Seth 2012, Buckner and Krienen 2013, Pascual-Leone 2006, Posner 1994, 2012, Striedter 2006).[13]

So far in this chapter we have discussed various forms of attention mainly with reference to behavioral and psychophysical studies in cognitive psychology. But these systems have also been studied extensively in the neurosciences, and evidence is found for distinct neural structures that support these different types of attention in humans (e.g., Kastner and Pinsk 2004, Kastner and Ungerleider 2000, 2001, Maunsell and Treue 2006, Yantis and Serences 2003) and animals (e.g., Matzel and Kolata 2010, Zentall 2005). For example, brain imaging studies on humans, as well as on nonhuman primates using more invasive single-cell recording procedures, have identified the hierarchical organization of visual areas that generally increase in complexity in terms of how the visual features are processed, from feature-based to object-based attention (for a review, see Kastner and Ungerleider 2000). Therefore, one can examine the neural systems supporting the various forms of attention to better understand the evolution of those functions. See figure 2.5 for a diagram of the brain areas associated with the various forms of attentional processing.

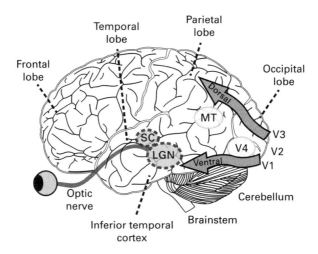

Figure 2.5
Brain areas associated with attentional processes. The major areas discussed in section 2.2 are labeled here, including the dorsal and ventral processing streams that support bottom-up attention. The flow of visual information moves from the retina, through the optic nerve, to the superior colliculus (SC) and separately to the lateral geniculate nucleus (LGN) of the thalamus. From the LGN, information is relayed to the visual cortex (V1–V5) and then to higher cortical areas. In terms of feedback processes, these can occur from any higher to lower levels (e.g., from V3 to V2), with 'executive' top-down control functions primarily originating within the frontal cortex. As a rough approximation, the evolutionarily newer areas of the brain include the frontal lobe with older systems closer to the brain stem.

Posner and Petersen (1990) describe three separate neural systems of varying functional complexity that are individually responsible for alerting, orienting, and target detection or executive functions. Different features, as well as focused versus distributed attentional strategies, also tend to activate different neural systems (Corbetta et al. 1991). Other studies in support of brain specialization have identified separate underlying neural systems that are activated when a test subject performs certain tasks that require different kinds of attention, such as motion-sensitive areas in the middle temporal (MT) region when tracking moving objects, or suppressive interactions (inhibition) from the extrastriate areas V2 and V4 when searching through sets of objects for a specific target (Bundesen, Habekost, and Kyllingsbaek 2005, Desimone and Duncan 1995, Kastner and Ungerleider

2000, 2001, Luck et al. 1997, Mangun 1995, Posner 1992, Wyart and Tallon-Baudry 2008). The proposed structure of these systems has generally held up over years of research, with some new evidence for additional attention networks for self-regulation and self-control that also contribute to attentional processing (for a review, see Petersen and Posner 2012).

One of the more primitive brain areas related to visual attention includes the superior colliculus (SC), which is located in the brainstem with connections that originate from the retina but are independent from the dorsal and ventral processing streams in the visual cortex. The SC contains a retinotopic map whose neurons are spatially organized like those on the retina. It codes the saliency of features and is involved in visual orienting with different layers for sensory processing, which produces motor responses such as eye movements. Top-down and bottom-up signals converge on the SC for making fast eye movements, or saccades, via many lateral connections that also can produce inhibitory signals (Dorris, Olivier, and Munoz 2007, Meeter, Van der Stigchel, and Theeuwes 2010).[14] Auditory stimuli also are processed by the SC, but faster, and producing weaker signals, than visual stimuli.

This more basic brain area, with the saliency maps it contains, sends saliency signals to higher cortical areas. These saliency maps support simpler, rudimentary forms of attention orienting. In terms of feature processing, the SC influences the feature maps that are located in the occipital lobe, which appear to have evolved in primates to process more complex features, such as complicated motions or the configuration of features that comprise faces. Such higher cortical areas also have evolved to modulate what is processed into higher-level cognition.

Many of the low-level, bottom-up visual processing systems for feature-based attention are organized in the visual cortex (see Hubel 1995, 93–125, Koch 2004, 49–86). Information from the retina flows through the optic nerve into the lateral geniculate nucleus (LGN) of the thalamus, before reaching the primary visual cortex (V1). From there, there are two major pathways, the dorsal and the ventral, also referred to as the 'where' and the 'what' pathways respectively (Kastner and Ungerleider 2000, Mishkin, Ungerleider, and Macko 1983, Ungerleider and Haxby 1994). The dorsal 'where' pathway (V1 → V2 → V3 → MT) tends to process information about motion and the location of objects; area MT is especially sensitive to motion and higher-contrast stimuli (Tootell et al. 1995). The ventral 'what' pathway (V1 → V2 → V4 → inferior temporal cortex) processes object feature information, including color, shape, and orientation. For example, color information, which is initially picked up by cones in the retina, is processed in area V1 and especially through area V2 and the LGN (for reviews, see Gegenfurtner 2003, Gegenfurtner and Kiper 2003). Orientation-specific cells also in V1 can detect orientation, contour, and curvature, which are important for determining shapes and objecthood (Beaudot and Mullen 2006, Ferster and Miller 2000, Hubel and Wiesel 1959, 1962).

There is another view, however, about these separate processing streams. Mel Goodale and David Milner describe the ventral stream as supporting 'vision for perception' and the dorsal stream as supporting 'vision for action' (Goodale 1998, 2011, Goodale and Milner 1992, Milner and Goodale 1993, 1995, 2008). By this account, the ventral pathway provides the contents of visual experience and is processed separately from the dorsal pathway, which provides visual information used to execute motor commands (which we discuss in section 2.3). This functional difference suggests that visual perceptual experience does not require all of the precise and immediate information necessary for executing actions, such as reaching to pick up an object (which requires knowing its size and location). Often, the visual information for successful actions travels only in the dorsal pathway and serves its purpose without reaching awareness. This anatomical organization, where information from the dorsal processing stream need not reach conscious awareness in order to affect behavior, would account for the often observed dissociations between attentional processes and consciousness.

For spatial attention, the supporting neural systems are thought to be located in the precentral sulcus, intraparietal sulcus, and lateral occipital sulcus, with significant modulation by the frontal cortex (Beauchamp et al. 2001). Additionally, there is some evidence for differential, though partially overlapping, networks within those regions that are responsible for bottom-up (exogenous) and top-down (endogenous) shifts of spatial attention (for a review see Chica, Bartolomeo, and Lupiáñez 2013). The spotlight model has some support via studies that identified activations as early as V1 during the shifting of attention, but it is also modulated by higher cortical areas, such as the ventral occipitotemporal cortex (Brefczynski and DeYoe 1999). Evidence also suggests that large portions of the cortex are involved with the shifting of attention, even in simple tasks (Ward 2013). The act of shifting attention from one spatial location to another is induced by signals from the posterior parietal cortex, part of the dorsal pathway (Yantis et al.

2002). The superior parietal cortex, which supports attentional capture, is also related to spatial shifts of attention (de Fockert et al. 2004), indicating a relationship between systems underlying spatial and feature-based attention (which supports models like feature integration theory). This illustrates how widespread and inter-related the systems supporting spatial shifts of attention are within the brain.

Bottom-up processes, like attentional capture, are influenced by neural networks that operate in a winner-takes-all fashion. There are also other similarities in the neural mechanisms for feature-based and spatial attention, particularly in relation to the modulation by spatial attention in areas associated with specific types of features (Maunsell and Treue 2006). These bottom-up systems are generally feedforward processes, but also include feedback neural structures that assist with top-down visual selection. A neural signature signal, the 'N2pc,' is related to the attentional selection of target features by suppression of the surrounding information (Hickey, McDonald, and Theeuwes 2006). Feedback and recurrent processes have been identified as being especially important for explaining how top-down modulation occurs, and may be related to conscious awareness (Lamme 2006, Pollen 2003). Again, we can see how these different forms of attention do not rely on completely independent brain structures, since there is plenty of overlap among their networks.

Further up in the visual cortex, the MT area processes motion information (Newsome and Paré 1988) while area V4 processes orientation information into object shapes (Haenny, Maunsell, and Schiller 1988). The global grouping of contour information (i.e., the Gestalt notion of 'good continuation') is important in the formulation of shapes and objects (Geisler et al. 2001, Kovács 1996) and relies on cortical processes that are responsible for global contour integration (Tanskanen et al. 2008). That is, such processes assemble holistic representations of objects. This integration process has been shown to be important in addressing how the visual system can extract shape information from images based on low-level processing mechanisms (Marr 1980). These sorts of higher-level systems are crucial for object-based attention.

The studies we have been citing suggest that attention does not operate only as an early selection mechanism, as some theories claim (e.g., Vecera and Farah 1994), but also can select objects in addition to locations, depending on the task at hand. Evidence supporting object-based attention comes from studies that reveal separate systems for spatial processing and object processing (Haxby et al. 1991, Yantis and Serences 2003) and show that attention tends to operate on whole objects rather than individual elements of objects. For example, one study differentiated object-based attention from spatial attention by testing for, and detecting, a within-object advantage in a subject (D.F.) with a lesion in the lateral occipital (LO) area in the ventral stream, which plays an important role in shape perception (de-Wit, Kentridge, and Milner 2009). Other studies show that even when only one feature is relevant for a task, all the attributes of the object with the target feature are automatically enhanced via corresponding activations of brain regions (e.g., O'Craven, Downing, and Kanwisher 1999), supporting previous behavioral studies that found the same sort of results (e.g., Duncan 1984, Egly, Driver, and Rafal 1994).

Object-based attention integrates information from various sources, such as individual features or motion information. Consequently, object-based attention is associated with a larger area of neural activations related to this processing (Roelfsema and Houtkamp 2011, Wannig, Stanisor, and Roelfsema 2011). Although object-based attention is related to spatial attention, some distinct regions of brain activity are associated with each; for example, anterior and centroparietal regions are associated with spatial attention, while activity within occipitotemporal areas and the posterior parietal cortex is associated with object-based attention (He et al. 2008, Shomstein and Behrmann 2006). Object recognition must involve higher systems as information is processed from the lower-level visual cortex to higher-level areas (e.g., inferior temporal cortex), finally mediated by the anterior fusiform gyrus; a continuous level of related cortical activity is associated with conscious awareness (Bar et al. 2001).

To summarize, the mechanisms for feature, spatial, and object-based attention have generally evolved from the lower-level areas of the brain and thus are closely related to bottom-up processes. One could describe attention to features as the first instance of a selective information processing system. Feature-based attention inherently requires some level of spatial processing, which then evolved into more advanced shifts of spatial attention (e.g., covert attention) as more complex object-based representations formed along with the capacity for multiple object representations in working memory. To organize all this information, the brain required more complex top-down modulation systems. To support a higher-level voluntary and top-down attention, studies have identified various brain regions considered to be more evolutionarily advanced, including the prefrontal cortex, which is associated with 'executive functions' (Baddeley and Della Sala 1996, Petersen and Posner 2012). Parietal and frontal regions in the cortex seem to participate in attentional control by initiating and maintaining attentive states (Yantis and Serences 2003). Additionally, the sense of 'self' is thought to rely on complex connections among the neural hierarchy of systems that support these various forms of attention (for a review, see Feinberg 2011).

Goal-directed top-down attention, nevertheless, is constrained by low-level computations in the visual system (Yantis 2000), highlighting its reliance on bottom-up processes. Earlier models of selective attention propose that objects in a visual scene compete for selection, and this competition is biased by bottom-up processes as well as top-down modulations, including task goals (Desimone and Duncan 1995). Such findings further indicate the importance of understanding how bottom-up and top-down processes interact (for a review, see Corbetta and Shulman 2002). In fact, recent studies suggest that feedback and recurrent processes are especially important for understanding top-down attention and may even describe how conscious attention arises (LaBerge 1997, Lamme 2006, Pollen 2003, Theeuwes 2010). From these studies on the organization of the brain (illustrated in figure 2.5), it is clear that different forms of attention evolved separately and became integrated at different stages in the development of our species.

As for the mechanisms that support attentional control, some researchers propose that attention relies on brain wave oscillations, that is, specific temporal frequencies of recurrent brain activity. For example, alpha and gamma waves seem to be involved in highlighting salient visual stimulus information during attentional capture (Jensen, Bonnefond, and VanRullen 2012), inhibiting irrelevant information or sustaining attention to relevant information (LaBerge 2002), and other functions that still need to be clearly identified (for reviews, see Klimesch 2012, Singer 2013). Oscillatory functions may also provide a way to unify and integrate visual information that supports conscious experience (LaBerge 2001). The idea is that when these oscillations are in sync, the neural activity is strong enough to produce conscious experience (whatever that may actually be). While this provides a theoretical perspective on how conscious awareness may arise, it does not improve our understanding of awareness, because these oscillations can occur and integrate information regardless of whether or not that information enters consciousness.

Furthermore, one must also consider the influences of neurochemical factors when trying to understand how attention and consciousness are supported by brain processes.[15] For example, serotonin is involved in visual information processing, and in cases where there are deficits in serotonin systems, as in lifetime drug users, there are deficits in sustained divided attention tasks and in higher-order visual skills such as perception of self-motion or spatial integration of complex visual scenes (Parrott 2013). Other research suggests that dopamine may also serve an important role in supporting consciousness by modulating neural activity (see Palmiter 2011). Additionally, clinical studies have examined various conditions, such as schizophrenia, autism, and neurochemical imbalances, in terms of how they affect attention, consciousness, and self-awareness (e.g., Bacon et al. 2001, Baron-Cohen 1999, Daprati et al. 1997, Lombardo et al. 2010, Sass and Parnas 2003, Spencer et al. 2011).

Much insight could be gained from examining deficits in attention. It is possible that such deficits produce a qualitatively different conscious experience, due to problems of integration, as well as preventing "typical" selection processes of attention. There may also be a relationship between irregular oscillations of brain waves and cognitive disorders such as schizophrenia or attention deficit (Calderone et al. 2014). Such clinical studies can reveal important information about the organization of systems that support attention, such as working memory, as well as the complexity of factors that contribute to proper functioning of these systems. These examples of clinical studies are promising and require further development.

The main point we would like to make with this brief look at the neural correlates of the different forms of attention is that the development of attention is described by an evolutionary story. From its beginning as a function of low-level feature detectors and spatial filtering mechanisms, attention developed into shifting mechanisms that can interact with more complex object-based representations. The importance of considering the neuroscience behind attentional processing is to identify how different forms of attention may operate independently from one another, or at least may be accomplished by a variety of systems. Clarifying the dissociations among the different forms of attention can lead to insights about consciousness, especially if consciousness is associated strictly with a more complex integrative process. For example, some central brain areas, such as the claustrum, are connected to many cortical areas and may point to a mechanism for directing and integrating information into a "conscious field" (Crick and Koch 2005). Many researchers, including Victor Lamme (2006), propose studying the neural correlates of consciousness through an empirical approach that relies heavily on studying attentional systems. There is much to be learned about attention and consciousness within cognitive neuroscience, and this area will continue to provide insights as the tools, techniques, and approaches are improved. The best we can say now is that the distinct systems that have been identified suggest evolutionarily separate adaptations that grew from basic visual information processing systems to more complex, sustained, object-based attention systems.

2.3 Attention without Conscious Awareness

Although we will go into more detail about consciousness in the next chapter, we should mention some studies that have identified the presence of attention without conscious awareness. As stated earlier, we refer to conscious awareness as the reportable qualitative characteristics of a perceptual experience. A conscious percept, in psychological terms, is associated with the ability to sense and reflect on perceptual information obtained from several modalities. For example, while our friend Lloyd, an established science-fiction writer, works on his computer, he can be conscious of the classical music playing in the background while also being conscious of his neck tension and of the typed words appearing on the computer screen. In this example, Lloyd is focusing primarily on the thoughts he is trying to articulate and how they are transmitted to the computer by scanning the visual information on the screen; he is minimally conscious of his fingers typing, his sore neck, or the background music—until perhaps a favorite melody comes on, and then he may be distracted from writing for a bit. This familiar example illustrates how attention continually processes information without that information always being the center of conscious awareness (as in the cocktail party effect mentioned earlier).

As a professional writer, Lloyd also experiences an effortless form of attention while he types out his stories. The motor commands associated with typing require some crucial information processing elements, like knowing the location or layout of the keyboard characters and where to place his hands. But being able to type in an effortless way requires Lloyd to be an experienced typist who no longer needs to hunt and peck for individual letters. In fact, good typists are likely to type more slowly if they deliberately think about what they are doing physically as they type. So while Lloyd may have had to be effortfully attentive when he was first learning to type in his youth, now he no longer consciously attends to the keyboard but rather more so to his thoughts that are transmitted to the computer screen. The information needed to locate the keyboard and process the motor commands is subject to forms of attention, but these types of attention are not necessarily conscious. This phenomenon is also experienced by musicians when they play music, as our multitalented Lloyd also does as an expert pianist. It is from such instances of effortless attention that the experience of flow emerges, which we will discuss in more detail in sections 3.5 and 4.3.

From various studies we know that many attentional processes do not reach consciousness, and yet the information they process can still affect behavior, decision making, and motor actions (e.g., Cohen et al. 2012, Dehaene and Naccache 2001, Desmurget et al. 2009, Kühn and Brass 2009, Mulckhuyse and Theeuwes 2010, Norman, Heywood, and Kentridge 2013, van Boxtel, Tsuchiya, and Koch 2010, Wegner 2003). Such studies indicate that perceptual decisions and motor actions are executed before one becomes conscious of the decision or intention to make the action, and that different neural correlates may support the subjective experience (Filevich et al. 2013). So in the case of Lloyd, although he is experiencing the transmission of thoughts as he writes his story, the attentional processes and motor commands to perform the typing actions occur before Lloyd subjectively "thinks" of the words in the story (a phenomenon that may be related to effortless attention).

In the case of visual illusions, attention can correctly process visual information to support actions even when the conscious perception of the stimulus does not match what is actually out there. For example, the Müller-Lyer illusion, where the perceived length of a line is influenced by surrounding information (i.e., arrows pointing in different directions), affects conscious awareness in terms of how the observer reports the lengths of these lines. Yet, this illusion does not influence motor control, such as grasping behaviors made to one of these lines (Stöttinger and Perner 2006). This is an example of the mismatch between the information used for perception and that used for motor actions.

Phenomena related to "failures" in attention, such as in blindsight, change blindness, or inattentional blindness, also describe occasions when attention does not provide correct information for awareness. The fact that such failures may be due either to the relevant information not being detected by low-level sensory receptors or to higher-level failures that prevent information from entering conscious awareness both exemplifies the complexity of the systems that make up the broad mechanism of 'attention'— all of which have the common goal of selecting certain perceptual information for higher cognitive processes—and provides support for a dissociation between attention and consciousness. A primary insight from these studies is that attention tends to be concerned with connecting mental representations (and other aspects of cognitive processing) with objects in the world, and it seems to be analytic and selective in nature—something like an information processing gateway.

Consciousness, on the other hand, seems to be a more integrative process. The hypothesis that consciousness is related to highly integrative processes has been supported by evidence from various research studies in neuroscience (e.g., Dehaene, Sergent, and Changeux 2003, Di Lollo, Enns, and Rensink 2000), and examining recurrent feedback networks in the brain is an emerging direction for understanding consciousness (LaBerge 2005, Lamme 2004, Raffone and Pantani 2010). Some have developed biological models aimed at incorporating the computational underpinnings of such highly integrative processes, guided by the theoretical concepts discussed above. For example, Giulio Tononi (2004, 2008, 2012) argues that conscious experience is related to the integration of various forms of sensory information, some of which may remain unconscious, or do not reach the threshold that makes them conscious, and nevertheless help constitute consciousness. That is, the degree of consciousness may be seen as directly related to the degree of integrated information. This possibility is supported by various behavioral and neural studies on the correlates of consciousness (for reviews, see Koch and Tsuchiya 2007, Mudrik, Faivre, and Koch, 2014).

Some of the important empirical studies on consciousness have used established experimental research methods on unconscious perception and unconscious cognitive processing. Baars and colleagues (Baars 1988, Baars, Banks, and Newman 2003) compare conscious processes with well-known unconscious processes that have been the subject of empirical studies, like inattentional blindness and the implicit attention involved in priming studies. Several researchers have studied cases where a subject is unable to report on a stimulus but can act on it in ways that indicate selective attentional processing must have occurred to enable that action (for reviews of such studies, see Chun and Marois 2002, Lamme 2006).

In the case of blindsight, for example, the subject cannot report seeing or experiencing the features of a target object, yet when asked to reach to it or act on it in some way, they accomplish the task in a manner that demonstrates a proper processing of stimulus features (Brogaard 2012, Kentridge 2012, Kentridge, Nijboer, and Heywood 2008, Weiskrantz 1996). Blindsight occurs when there is damage to area V1 that prevents information from reaching the higher levels that would allow conscious perception. Information, however, still reaches the SC (which is directly connected to the LGN), which then provides the motor system and cortical areas with the information relevant for performing actions (Weiskrantz 1996). Similarly, some people who are diagnosed as clinically blind are able to navigate around obstacles (de Gelder et al. 2008), which can be taken as support that 'vision for action' can be processed separately from 'vision for perception' (for reviews see Brogaard 2011a, Kentridge 2012).

A more common phenomenon, change blindness, describes instances when an observer is unable to notice an obvious change in a scene until a certain point when the change enters awareness, even when the observer knows they should be looking for a change (Rensink 2002, Simons and Rensink 2005). This has been tested using the flicker paradigm (Rensink 2009, Rensink, O'Regan, and Clark 1997), which repeatedly presents to an observer two versions of the same image in succession, with one significant item removed from one of the images, and measures how long it takes for the observer to identify the missing item. In one experiment, for example, two pictures of a café scene are shown with one of the tables missing in the second photograph. It usually takes some time before the observer notices the missing element, and when one finally does notice the missing item, it seems surprising that this obvious change has been missed. Figure 2.6 shows other similar examples. Change blindness, since it suggests that attention is necessary in order to consciously perceive changes in visual information, can be used as a tool to study conscious perception and attentional allocation mechanisms (Rensink 2008). This change blindness, however, is less likely to occur with stimuli such as animals and humans, suggesting an evolutionarily adapted mechanism that monitors such information (New, Cosmides, and Tooby 2007).

Figure 2.6
Example of stimuli that would be used in a change blindness task (something from the top image is missing in the bottom image). In this experimental task, each image pair is viewed alternately, and the observer is required to report any change between the two scenes. The tendency not to notice change, or change blindness, occurs often, even for large objects that seem obvious once they are finally noticed.

Related to change blindness is inattentional blindness, another instance of the failure to notice an object because attentional processes are heavily loaded or directed toward other aspects of a visual scene, even when the object spatially overlaps the areas to which attention is directed (Mack and Rock 1998, Most et al. 2005, Rock et al. 1992). A popular example of inattentional blindness is the test wherein observers watched a video of a group of people passing a basketball to each other and were asked to count the number of passes made to a subset of players. While performing this attention-demanding task, a majority (nearly three-quarters) of observers did not notice a person in a gorilla suit walking through the scene (Simons 2000, Simons and Chabris 1999). While attention was focused on ball passes and on maintaining a running count of the target passes, the perception of something irrelevant to the task did not enter conscious awareness—even something as contextually surprising and salient as a gorilla on a basketball court. Such studies illustrate the level at which selective attention must be engaged in order to contribute to a reportable conscious experience of visual objects (Lavie, Beck, and Konstantinou 2014), and indicate that some systems of spatial attention are related to those of attentional capture (Most 2010, Most et al. 2001).

The temporal limitations of attention are often studied under the *attentional blink* paradigm, which identifies failures of attention during a window of time after the detection of a target (for reviews see Dux and Marois 2009, Marois and Ivanoff 2005). These studies typically use a rapid serial visual presentation (RSVP) task, which presents stimuli in rapid succession with varying gap durations between stimulus presentations. When a specific target appears that must be reported, there is a window of time in which the processing of a second target fails, usually within 200 to 500 milliseconds after this first target. The failure to detect the second target in the RSVP stream is called the 'attentional blink' (Broadbent and Broadbent 1987, Raymond, Shapiro, and Arnell 1992). Attentional blinks may be due to interference from top-down signals that serve to increase the competitive strength of the first target while hurting the second target (Hommel et al. 2006). Also, neuroimaging studies indicate that the frontal cortex is involved in conscious reporting, and signals that fail to reach this area are correlated with the attentional blink (Marois, Yi, and Chun 2004). Such studies on the failures of attention suggest that conscious awareness relies on activations in the frontal cortex, at least in cases related to the attentional blink.

Even during the "normal" operations of the other forms of attention, not all processed information reaches consciousness. Some studies on object-based attention have shown that it occurs even when one has no conscious awareness of that object (Chou and Yeh 2012, Mitroff, Scholl, and Wynn 2005). As discussed above, previous object-based attention studies have shown a within-object advantage for processing features in an object, even for nontarget (task-irrelevant) features. In a recent study that examined whether or not awareness was necessary for object-based attentional selection (Norman, Heywood, and Kentridge 2013), test subjects exhibited a within-object advantage in a cuing task (similar to that shown in figure 2.3) under conditions where the observers were not consciously aware of the presence of the objects due to the use of a continuous masking paradigm. Even when not aware of the cuing object, observers were faster to respond to a target when it was presented on the same object that had been presented before the target's appearance, compared to when the targets appeared at the same distance on different objects or in locations with no objects. This finding supports the idea that attention can select and process object-based information automatically, even when observers cannot report seeing the objects (i.e., when the information does not reach awareness). One explanation for this phenomenon is that the mask may prevent the attended stimulus from reaching the threshold that moves it into awareness; another is that the mask prevents the recurrent processing that may be responsible for awareness (see Fahrenfort, Scholte, and Lamme 2007). So, while object-based attention operates independently and can affect behavior, higher-level processes related to awareness do not necessarily occur along with it.

Attentional processes can also be concerned with producing an integrated and stable representation, which often requires the filtering of irrelevant information. Studies on *visual stability* address the question of how it is possible to have a stable percept of the visual world given all the disruptions in information processing caused by the many blinks and eye movements that we make (for reviews see Cavanagh et al. 2010, Mathôt and Theeuwes 2011, Melcher 2011). We tend to have a stable representation of our visual world even though we constantly make fast saccadic eye movements that produce irrelevant motion information. For example, as you move your eyes around the room you are in, you rarely notice the motion blur caused by your eye movements, although you can make yourself notice it if you try (not a recommended practice for those susceptible to motion sickness). The visual system suppresses this irrelevant motion information and also integrates the changing streams of visual input (occurring on different parts of the retina during eye movements) into a stable percept. A lot of information is processed by attention and integrated in the background to help produce stable perception, but it does not always reach conscious experience.

Many studies have been conducted on visual stability and on the suppression of the irrelevant visual information, called saccadic suppression or blink suppression, that helps produce it (e.g., Bays and Husain 2007, Ridder III and Tomlinson 1997, Watson and Krekelberg 2009). Despite the constant presence of such visual disruptions, we somehow remain conscious primarily of the information that produces a stable percept of the world.

This ability—to integrate seemingly disparate streams of information into a contiguous and stable representation while filtering out irrelevant information—can be thought of as an example of conscious attention. The significance of these suppressive and integrative processes of attention, and their relationship to consciousness, are not yet fully elaborated, but they are of particular interest because they directly affect what is consciously perceived.

So, what can we say now about the relationship between attention and consciousness? Are all forms of attention forms of conscious attention? Intuitively, this would seem likely to be the case, since attending to something would seem to require some conscious awareness of it. The empirical evidence, however, paints a much more nuanced picture. As we have discussed, attention operates at different levels with several varieties and functions such as preattentive, distributed, and focal attention. Since such a variety of layers are associated with attention, not all of these functions will reach conscious awareness. The information processed by attention need not correspond perfectly to what is perceived; different details are processed by the 'vision for perception' system and the 'vision for action' system, for instance. Additionally, the recent work addressing the relationship between consciousness and attention across different sense modalities raises a common question of whether or not attention is necessary for conscious perception, and includes arguments for both sides of that debate (see Tsuchiya and van Boxtel 2013). So while it seems that attention is necessary to process information that reaches the contents of awareness, it is not *sufficient* for consciousness (see Cohen et al. 2012). This conclusion, however, is contested by some theorists who argue that attention is both necessary and sufficient for consciousness, as for example, J. J. Prinz (2012), who claims that attention and consciousness are identical processes.

A further complication is introduced by the idea of 'creature consciousness': some attempts to identify the presence of consciousness in organisms characterize it as either intransitive or transitive. If such a distinction is accepted, then intransitive creature consciousness—what is it like to be an organism—may just reflect the qualitative character of wakeful states of such an organism (as opposed to sleeping or being comatose). But it would seem that transitive creature consciousness, in which the creature is not only consciously aware but is consciously aware *of something* in particular that is being perceived, must be mediated by forms of attention. So a further dissociation is introduced by such distinctions, with attentional mediation clearly involved only in perceptual forms of awareness, or transitive consciousness. Whether or not creature consciousness should be characterized as being some form of conscious attention is a controversial issue that we discuss in chapter 5.

Based on the studies we described, conscious attention accounts for only a portion of the processes that fall under attention. Although consciousness is influenced largely by focused attention, it also comprises cognitive events that are not attended to in the strictest sense; that is, it is not simply a focus of attention. Consciousness includes all the bits and pieces in the background, including your current comfort level (like adrenalin level, pain, or hunger), sensory information that may be tuned out (like background music, the movement of the train you are riding, or the accustomed noise of the cars passing by your apartment window), and the several streams of thought that might be contributing to your distracted state. All these cognitive bits make up consciousness and conscious experience, and clearly not all of these pieces are consciously attended, even though attention is processing them. Additionally, not everything has to enter consciousness in order to influence cognition or actions. Indeed, some researchers have argued that much of behavior occurs outside of voluntary conscious attention (e.g., Bargh and Chartrand 1999, Bargh and Ferguson 2000). It is also argued that although there are some neural overlaps in the mechanisms of attention and awareness, the experience of conscious awareness may be more closely tied to working memory systems (Lamme 2003, 2004).

The empirical work on visual attention may be our best link to the study of consciousness, even though the relationship is not always straightforward given the interactions of the various attentional mechanisms with other brain processes including memory and motor systems. Numerous studies provide support for the argument that many attentional processes occur outside of consciousness, and even suggest that some conscious experiences occur outside of attention, although the latter is harder to test. Evidence from neural studies points to dissociation between low-level and high-level forms of attention, supporting at least a Type-A CAD. By further examining and clarifying the relationship between consciousness and attention during conscious attention, we can gain a few insights about ways to study consciousness and develop methods to better understand it.

2.4 Forms of Conscious Attention

One of the challenges in the empirical study of consciousness is the task of explaining the nuances that describe the current theoretical understanding of consciousness at the neural level. For example, what would it signify if a specific pattern of neural activation was empirically confirmed to correlate with consciousness? Does it mean we can distinguish between the different types of consciousness that have been proposed by philosophers? Could it be that the pattern of activation is simply correlated with consciousness, but neither explains nor identifies the mechanisms of consciousness or how it corresponds to attention? Could activation patterns correspond to the *integration* of information but not to the *sources* of information on which the mechanisms of integration depend?

Despite the difficulties underlying the metaphysical understanding of consciousness, we believe that the progress on the experimental front has been substantive. By focusing on the largely unexplored implications of *conscious attention*, we can outline the general features of an adequate theory of consciousness that would successfully guide future empirical research. We will further develop this argument in chapter 4, after describing the current theoretical understanding of consciousness in chapter 3, but here we introduce some of our main ideas, emphasizing how they relate to various levels of the consciousness and attention dissociation (CAD).

Conscious attention can be defined as a phenomenal experience of perceptual information, directly resulting from attentional processing and thus available to cognition in a reportable manner. In a philosophical sense, conscious attention requires a "demonstrative awareness that one is attending to *that* object" (Wu 2011), which also entails voluntarily maintaining attention to a specific perceptually selected object. One important consideration when describing conscious attention is that it inherently requires the integration of multiple kinds of information into a coherent percept, either within a single modality or crossmodally. For instance, both the visual and auditory features coming from a car may be needed to determine that the fast-approaching object is in fact a car. A *global workspace model* of consciousness addresses such concerns, providing a theory by which to test the idea that different kinds of information can be made available in some unified manner for thoughts, decision-making processes, and action planning, that is, in a process that enables a 'global access' (e.g., Baars 2005, Raffone and Pantani 2010).

The integration of perceptual information is important for creating feature-rich visual representations such as the object files we discussed earlier. The ability to integrate information crossmodally is also crucial for complex behaviors that increase the probability of success at such vital tasks as foraging and hunting (Gallistel 1990a). In humans, consciousness is sometimes argued to be the necessary step in the integration of multimodal sensory information that is processed in isolation, such as sound and vision (e.g., Palmer and Ramsey 2012). The main role of consciousness, according to this view, is to broadcast the contents computed in a uniform, presumably conceptual, format, which serves an epistemic role since it provides access to contents across different modalities. This is a central claim of global workspace or global broadcast views of consciousness (e.g., Baars 1988, 1998, Dehaene and Naccache 2001). Of course, this perspective occasions the debate of whether or not animals possess consciousness, since they too require some sort of global, integrative mechanism.[16]

An extension of object files has been proposed by Hommel in *event file theory*, which incorporates multimodal sensory information and motor commands related to action planning (Hommel 2004, 2005, 2007b, Hommel and Colzato 2004, Zmigrod, Spapé, and Hommel 2009). In terms of sensory integration, we often take sound and visual information presented together to be part of the same event, as for instance when you hear and see a car approaching, and make plans to hurry out of the crosswalk that the driver seems to be ignoring. Even with asynchronous presentations of multimodal information—for example, when there is a 350-millisecond gap between the tone and the visual stimulus onset in a research setting—unconscious associations can be formed that allow asynchronous features from separate modalities to be bound together as part of the same event, even if they are not consciously perceived as unified (Zmigrod and Hommel 2011). In other words, background processes can be more generous in assuming what events sound and vision correspond to than is perceived within conscious awareness. This sort of feature binding does not necessarily require an awareness that it is happening, and the presence of a temporal asynchrony is often overlooked.

The importance of event file theory is that it addresses the ultimate purpose of the ability to attend to things in the world and represent them—namely, to act on them. The model describes a source of agency for the system of representing event information, which may be related to abilities such as tool usage. Although event file theory as proposed requires more development and testing, there must be some set of processes that allow the mind to incorporate information from the motor system into perceptual representations for action.[17]

In relation to conscious attention, object files and event files may provide the structures and contents for conscious experience, though not all their elements need to reach consciousness. Conscious attention is the culmination of various separate attentional processes, which are influenced by how the information attended is modulated via top-down and bottom-up mechanisms. Not everything that has been processed by the different forms of attention is made available to consciousness, but the elements that are reportable enter via conscious attention. So, then, what determines the information that enters conscious attention, and what role does conscious attention ultimately play?

Influences from top-down control via recurrent neural processes appear to be important for conscious attention. Lamme (2003, 2004) proposes that conscious experience and awareness, that is, conscious visual attention, relies on a recurrent processing that modulates or selects information that enters the visual cortex initially through the feedforward sweep (FFS) stage. The FFS is the earliest level of activation that is related to feature-based attention and other low-level processes. Recurrent processes, the sending of signals back to the low-level areas, can occur as soon as information reaches a higher area during the FFS, and these recurrent signals appear to be related to awareness (Lamme 2004). Lamme argues that it is more productive to identify consciousness in such a process of connectivity and transfer of information, as opposed to identifying the brain areas specifically responsible for consciousness.

Another possible role of conscious attention is to help resolve ambiguity in images that may have multiple interpretations, like the duck-rabbit or the Necker cube in figure 1.2. The importance of such 'contrasts' is discussed by Block (2010) as a way to describe how attention changes phenomenal awareness. Various studies have demonstrated such effects, where directing covert attention toward a target tends to increase the perceived intensity of the target (e.g., Carrasco 2009, Carrasco, Ling, and Read 2004, Carrasco and Yeshurun 2009). These studies also indicate that both overt and covert attention can increase spatial resolution on the attended location or object, so that more information about the visual features is encoded. Whether or not attention is focused on a target, therefore, will inevitably change how that target is perceived.

Self-awareness is another aspect of conscious attention that must be considered. A representation of the self may rely on signals from the thalamus and may be integral to conscious attention (Bruya 2010, 12). In the *triangular circuit theory*, proposed by David LaBerge (1997, 2001), it is argued that consciousness arises through the activations of various areas responsible for attentional selection, enhancement, and control, together with a representation of the self in relation to the objects of attention. The nature of self-awareness, then, becomes an important issue, even though not all forms of conscious attention involve an explicit reference to the self. (It may be there *implicitly* all the time, but such a thesis also demands clarification.) Triangular circuit theory also implies that attention is necessary for consciousness; that is, one cannot be aware of something unless attention has operated.

Thus, the 'self' may be more important than the current theories about attention might suggest, given that reference to the self is rarely mentioned in studies on attention. An egocentric frame of reference, intentionality, and a sense of agency are inherent in the ability to plan and perform actions. But since the nature of subjectivity is problematic, its systematic study has largely eluded researchers and so we have little to say about it in this review. (We will return to the issue in chapter 4.) One problem with this perspective, however, is that if this proposal is interpreted in a species-specific manner, a sense of self and intentionality may be only human characteristics, and thus only our species could have conscious attention (unless we consider egocentric frames of reference as a fundamental form of self-awareness that is present in animals).

So what is conscious attention? From our review of the literature, we take conscious attention to be awareness of a limited set of information derived from various attentional processes—be it objects, or their features, or other aspects of sensory information. Reporting the contents of conscious attention is typically how one determines its presence, but this could be problematic, since self-report is not the ideal way to empirically test theories. Before we can further examine conscious attention in chapter 4, we need to discuss the various forms of consciousness, which we do in the

Forms of Attention 77

next chapter. The main point to make now is that much more attentional processing occurs than what reaches consciousness, especially from the perspective that attention serves action. Conscious attention contains only a subset of the cognitive states related to attentional processes, and in the end, given these dissociations, one cannot justifiably reduce consciousness to attention. That is, there cannot be a one-to-one correspondence between attentional processes and the contents of conscious awareness. Any overlap, however minimal it may be, could be considered a form of conscious attention. Although more empirical work is needed to address this issue, the current findings give us good reason to believe that attention and consciousness are largely dissociated.

What, then, can be said about the relation between consciousness and attention? In terms of the CAD levels delineated in chapter 1 (see figure 1.1), given that there are separate forms of attention that could reasonably be supported by separate neural systems, we believe that evidence will indicate at least a Type-B CAD. Even within attentional processing, independent neural pathways have been shown to provide separate information for perception and action, such that actions can be performed successfully even when the perception of the stimulus is incorrect (as in the cases of visual illusions or blindsight). These findings have direct consequences for theories of consciousness, since a dissociation within the types of attention, especially in terms of what reaches conscious attention, will weaken any identity claims between consciousness and attention.

For the relation between consciousness and attention to be a Type-C CAD, there would have to exist several types of conscious attention that are separate from each other. That would be supported by the distinctions between distributed spatial attention and object-based attention, between covert and overt attention, and between effortful and effortless attention (where each of these distinctions can have their own form of conscious attention). The challenge ahead is to determine which forms of attention are more relevant for consciousness. It may seem intuitive to favor effortful attention and object-based attention; since they require more cognitive resources for integration, they should be more related to conscious experience. Yet, the instances of attention that do not reach consciousness show us that the relationship between consciousness and attention is even more complicated. Nevertheless, what seems unambiguous from the evidence so far adduced is that the varieties of attention can be dissociated at some level. So we have to reject the identity thesis for the relation between consciousness and attention.

2.5 The "Hard Problem" of Consciousness and the Easier Problem of Attention

As shown by the many studies described so far—and we have reported on only a fraction of the published studies—attention is typically examined in an empirical context. Cognitive psychologists now use various computer-based tasks that test observers on their ability to notice or remember target features or objects. When the experiments are well designed, these behavioral and psychophysical measures can be quantified in a fairly straightforward way. Also, neuroscience methods have examined attentional systems to understand the organization of the brain and to test if distinct structures are involved, and those methods will only improve with technological advances (e.g., see Cavanna and Nani 2008). Thus the "problems" of understanding attention have become empirical questions, and therefore addressable by thoughtful experiments in psychophysics or neuroscience.

The study of consciousness, however, was traditionally in the philosopher's domain. Given the subjective nature of conscious awareness, the empirical study of consciousness is a notoriously more difficult problem to address. Nevertheless, cognitive psychologists are increasingly interested in addressing the problem of consciousness empirically, often through studies on attention like those we have discussed above. In particular, the number of studies that apply neuroscientific approaches to the study of consciousness and attention has grown over the last 20 years, and such studies should continue to clarify the relationship between consciousness and attention (e.g., Cavanna and Nani 2008, Crick and Koch 1998, 2003, Dehaene et al. 2006, Dehaene and Naccache 2001, Di Lollo, Enns, and Rensink 2000, Koch 2004, 2012b, Koch and Tsuchiya 2012, LaBerge 1997, Lamme 2006, Posner 1994, 2012, Raffone, Srinivasan, and van Leeuwen 2014b, Taylor 1997, 2003, 2007, van Boxtel, Tsuchiya, and Koch 2010). Subsequent researchers have the task of discerning which lines of inquiry are most promising and which might be on the wrong track. Here we summarize some of the primary unanswered questions in need of empirical work.

Is attention required for consciousness, and vice versa? The literature we have reviewed includes cases of attentive processing that do not reach conscious experience (e.g., Dehaene et al. 2006), and some argue that visual conscious awareness does not necessarily result from attentional processes in the early visual cortex (Koch and Tsuchiya 2012).[18] Whether or not attention is absolutely necessary for consciousness continues to be debated (Marchetti 2012, Mole 2008, Taylor and Fragopanagos 2007, Taylor 2013, van Boxtel, Tsuchiya, and Koch 2010), particularly with respect to some of the more primitive forms of attention, like spatial attention (Hsu et al. 2011, Wyart and Tallon-Baudry 2008). Of course, many forms of attentive processes, as well as their integration, are required to provide the contents of consciousness, but this process needs to be detailed more rigorously. Our main argument here is that because of the different forms of attention and the presence of attentive processing outside of conscious awareness, we believe that a dissociation exists between consciousness and attention, and this perspective must be further explored.

Is the low-level versus high-level distinction a useful one? There are apparent "levels" of processing within the broad area of selective attention, as outlined above. Attention is often described as being bottom-up or top-down, in accordance with this processing dichotomy. But what does that truly tell us about consciousness? One can argue that the bottom-up forms of attention are more primitive, and are present and fundamental to a range of organisms, not just humans. It seems likely that with the evolution of a more complex organization and increased interactions among brain regions, a more advanced form of attention developed to enable the willful, top-down control of selective processes. This classification is not perfect, as shown by some recent studies, but it has helped organize our understanding of selective processes and their functions. Whether this distinction will continue to further our understanding of attentive processes is debatable, since the different "levels" are so intimately interrelated; but the top-down versus bottom-up contrast may still provide a useful framework for the evolutionary aspects of attention. This distinction also sheds light on studies on integration, executive control, and global workspace models (e.g., Baars 2002, Dehaene and Naccache 2001, Hommel 2007a, LaBerge 1997, Posner 1994, 2012, Tononi 2008), as well as the role of reentrant processes in producing conscious experience (e.g., Di Lollo, Enns, and Rensink 2000, Lamme 2003). Also promising are studies on low-level phenomena, like finding effects of unconsciously primed stimuli (Botta, Lupiáñez, and Chica 2014, Kiefer 2007), studies on how external stimuli can automatically enter consciousness in a reflex-like manner (Allen et al. 2013), and studies that find visual adaptation or suppression under conditions of binocular rivalry (Alais et al. 2010, Carter and Cavanagh 2007, Carter et al. 2005). In fact, binocular rivalry, which occurs when two different images are shown to each eye such that the viewer experiences only one of the images at a given time (and this percept alternates between the two images), may be particularly revealing about the way visual information is processed into conscious awareness (Blake, Brascamp, and Heeger 2014).

Is effortless attention associated with unconscious information processing, and if so, how? Understanding what makes attention effortful or effortless may provide insight into what enters conscious awareness. It would be easy to claim that consciousness characterizes the more effortful forms of attention, since it may serve the purpose of directing resources toward an effortful task or one that involves conflicting sources of information (Bruya 2010). Yet there are also forms of effortless attention that enter conscious experience, such as the experience of flow. So, what makes these forms unique? Research areas that can further our understanding of effortless attention include procedural learning (Willingham, Salidis, and Gabrieli 2002) or implicit learning (Cleeremans and Jiménez 2002, Goujon, Didierjean, and Poulet 2014), and how consciousness and attention may interact with such learning processes (Choi and Watanabe 2009, Custers 2011, Lovibond et al. 2011, Meuwese et al. 2013). Identifying how an experience can move from one of high effort to one that is effortless may illuminate the subjective nature of consciousness.

What is the role of memory in attention and consciousness? In their discussions of consciousness, many philosophers including Dennett (2005) have emphasized the role of memory. Even Friedrich Nietzsche (1887/1974) suggested that our highly evolved memory system is the defining feature that makes humans uniquely able to have conscious experience. The insights required to further our understanding of consciousness are unlikely to be found through attention research alone; working memory and access to long-term memory are crucial for maintaining and providing the content for sustained attentive and conscious states (Taylor 1997, 2007). Working memory is already being studied in terms of the 'central executive' and 'global workspace' models related to attention (e.g., Baars 2005, Lamme 2003, Posner 1994, Raffone, Srinivasan, and van Leeuwen 2014a). Memory certainly should not be overlooked in this endeavor, and many theorists are identifying key elements of memory systems and their relationship to conscious attention (for a review, see Markowitsch and Staniloiu 2011; we discuss the importance of autobiographical memory in chapter 4). It may be possible that the attentive maintenance of working memory contents and access to that information are the types of cognitive processes that directly correspond to conscious attention. Of course, this sort of question, too, requires empirical testing.

What is the role of language and the need to communicate in consciousness? Closely related to the use of memory systems in consciousness is the development of the capacity for language and the ability to communicate. Without memory, this would not be possible. Nietzsche (1887/1974) also proposed that consciousness may have arisen from the capacity for language. If consciousness can be shown to have evolved from the need for communication, it may be clearly related to the structures in attention and memory that enable language (for a related discussion, see Gordon 2009). Furthermore, some theorists argue that language is required for certain forms of consciousness, such as self-consciousness (Baker 2013, Dewart 1989, Neuman and Nave 2010). One problem with this account is that it suggests, counterintuitively, that infants may lack robust forms of consciousness.

What does the evolution of attention tell us about consciousness? Is there an adaptive advantage for consciousness? By clarifying the evolutionary influences on attention, we can better see how these relevant adaptations improved the fitness of various species, as more complex forms of attention produced more complex survival behaviors. How consciousness plays into this posited adaptation needs to be further explored (we discuss this in the book's conclusion). Is consciousness like attention, so that it can be described in evolutionary terms? We think advancements in the understanding of consciousness can be furthered by clarifying the relationship between attention and consciousness, especially in terms of whether consciousness serves a purpose independent from attention. For example, a recent study suggests that consciousness (not attention) is necessary for learning (Meuwese et al. 2013). Perhaps the integration involved in conscious attention enables learning, particularly for complex tasks, but it is questionable if conscious attention is a necessary causal mechanism for learning, especially in light of studies that support implicit learning and the learning that happens in animals—all of which provokes the question: Do animals have consciousness as well?

What is the relationship between phenomenal consciousness at the very low level, and the high-order consciousness that seems to require self-awareness at a metarepresentational level? Is self-awareness crucial for consciousness? Is it the defining characteristic of conscious experience? Studies on egocentric frames of reference in spatial navigation (e.g., Wang and Spelke 2002) may shed light on the origin of our more reflective experience of the self. It also seems that basic attention with some sort of self-attended activations provides the basis for conscious experience (LaBerge 1997, 2001), although this self-reflection is not necessary for attention. The question of subjectivity is a difficult one to study empirically, but since it is a central component of consciousness it must be addressed (Park and Tallon-Baudry 2014). Nonetheless it appears there is a dissociation between the systems for basic low-level attention (e.g., those involved in navigation) and the systems underlying attention that involves reflexive self-reference.

Indeed, another critical question is to what extent *should* 'subjectivity' shape the empirical study of consciousness? The answer will vary depending on who you ask. Some philosophers would argue that subjectivity is the central question motivating our drive to understand consciousness, while some scientists would argue that the nature of subjective experience makes it a very difficult phenomenon to study objectively with our current understanding and tools. Furthermore, some philosophers think that the first-person perspective is primitive and not susceptible to analysis or reduction (Baker 2013). For our part, we believe that like other bodily processes, consciousness (including its subjective nature) can and should be studied, tested, and observed from an empirical and quantitative perspective. Because consciousness is so very subjective, intuitively it seems more amenable to qualitative study. But while the subjective perspective is certainly useful to inform the empirical study of consciousness, it need not be the only approach (see Dennett 2005). By grounding the study of consciousness in visual attention, we can approach it using empirical tools that are already well established, can be described in evolutionary terms, and inherently address the question of subjective experience.

2.6 Summary

By describing functionally the varieties of attention and the manner in which they are controlled, we have shown that there are distinct systems to support the different kinds of attention. The inherent dissociation among the forms of attention corresponds to different functions, which can be described from an evolutionary perspective and are evident in the brain's organization; the more complex the attentive function is, the more association it has to newer cortical areas, along with more intricate connections to the other areas that support low-level attention. Conscious attention, which may serve to integrate features and information from memory, is a more complex form of attention that generally requires some sort of sustained activation of various brain regions. The kinds of conscious attention possible will be explored further in chapter 4.

An important conclusion at this point is that there are attentional processes that can affect behavior and actions—and thus are functionally selecting information, as attention typically does—but that do not reach the "reportable" space of conscious awareness. This dissociation has been tested through a variety of empirical studies, as we have described, with especially interesting findings in the case of blindsight. This dissociation between attentional processes and conscious awareness is important for our conclusion that while attention appears to be necessary in some way to provide the contents of conscious experience, consciousness need not to be present for attention to process information successfully. Additionally, methodologies used to prove that consciousness indeed occurs without attention have not been perfected, and the accounts of such instances are debatable. Before we closely examine the relationship between conscious awareness and attention, we must explore the various theories concerning consciousness. This will require us to delve into the philosophical literature, so be prepared to shift gears!

3 Forms of Consciousness

If we wish to determine the relation of attention to consciousness, at least some parameters of the two terms need to be established. In the previous chapter we looked in some detail at the various theories about attention, exploring the range and variety of cognitive functions associated with the term. After reviewing how our approach applies to theories of attention, particularly in light of the spectrum of dissociation and how it affects our understanding of conscious attention, we now turn to consciousness.

In this chapter we show that several theoretical considerations favor a robust form of dissociation between consciousness and attention. We systematically unify debates that have been largely isolated from one another. We argue that not only is dissociation a plausible view, but that the most severe forms of dissociation, those we termed Type-C, seem to be *entailed* by standard distinctions used in consciousness research. Our approach helps disambiguate terms that need to be reconciled in order to improve exchanges between theorists. Our account reveals several affinities among the extant views that have not been properly understood. The main conclusion of this chapter, consistent with our arguments in the book as a whole, is that many of the current debates point to the need for a more comprehensive theory of consciousness and attention, and that a dissociation between consciousness and attention is a crucial feature of such a theory.

The chapter is structured as follows. Section 3.1 provides a detailed account of the forms of dissociation entailed by the distinction between phenomenal and access consciousness, and by first-order and higher-order theories. Section 3.2 examines and clarifies the relationship between cognitive integration, the unity of consciousness, and access to information. It explains access to information in terms of a basic form of epistemic normativity, and suggests that epistemic agency need not be construed introspectively. Section 3.3 reviews evidence on consciousness without attention, analyzing it in the light of theoretical and neuroscientific considerations. Section 3.4 draws the implications of the previous sections for the debate on self-awareness, consciousness, and attention. Finally, section 3.5 presents the consequences of this chapter for the topics of conscious awareness, social influence, and effortless attention.

3.1 Consciousness: Phenomenal and Access, First-Order and Higher-Order

The distinction between phenomenal and access consciousness is very relevant for the topic of dissociation,[1] so we need to analyze it in some detail. This analysis shows that the dissociation between phenomenal and access consciousness is related to higher-order theories of consciousness, views on epistemic normativity, and accounts of the unity of phenomenal consciousness in ways that have not been noticed before, and we will consider the implications of dissociation for some higher-order theories. There are various possibilities for dissociation, depending how phenomenal consciousness is distinguished from access consciousness based on considerations concerning higher-order representation. An independent and equally important motivation to study higher-order theories is that one of the few proposals for the evolution of consciousness that is explicitly grounded in empirical findings is a higher-order theory. There are two versions of this theory, as will be explained in chapter 5. The conclusions from our review of first-order and higher-order theories support other types of dissociation between consciousness and attention, as will be discussed in the remainder of this chapter.

3.1.1 Phenomenal and Access Consciousness, and the Dissociation between Consciousness and Attention

Block (1995) argues that one must distinguish between two types of consciousness in order to avoid ambiguities in theoretical debates. One of them, access consciousness, is associated with the sense of 'awareness' that includes monitoring, selecting, and retrieving information. It is also associated with the control of sensorial and conceptual information and the use of such information for other cognitive tasks. Because of these roles, access consciousness seems to be deeply associated with selective attention, filtering, monitoring, and other functions of attention. A distinctive feature of access consciousness, however, is that it requires *global* access: if one is access-conscious of information, then the information is available for decision making, thought, linguistic report, and action. Global workspace theories endorse this definition of consciousness in terms of global access (see Baars 1988, 2005, Dehaene and Naccache 2001). Strictly functional accounts of executive control and its relationship to working memory are also intimately related to access consciousness (see Baddeley 2007, Gray 2004).

By contrast, according to Block and many others, phenomenal consciousness cannot be defined functionally, even if the information is globally accessible. Phenomenal consciousness is the qualitative character of an experience, and some philosophers have argued that it is irreducible to functional characterizations in terms of inputs and outputs (Chalmers 1996). Phenomenal consciousness is 'what it is like' to have an experience, rather than just having access to information. To use a traditional example, it is the redness that fully captures what seeing red is for an observer at a given time, irreducible to objectively measured color spectra, neuronal activity, and other physical properties. Methodologically speaking, this is a critical distinction. On the one hand, if the distinction between access and phenomenal consciousness is adequate, then standard methods to study the mind will never be able to fully explain phenomenal consciousness, because they are only capable of studying access to information.[2] On the other hand, if the distinction is inadequate, then consciousness almost certainly should be understood exclusively in terms of access consciousness, so that the study of conscious awareness can be the subject of empirical investigation.[3] These are important implications of Block's distinction, and it is no surprise that it has led to several versions of the following impasse: either phenomenal consciousness is the only type of consciousness that deserves the name 'consciousness,' or access consciousness is the only type of consciousness that can and should be studied.

One way of interpreting the distinction between access and phenomenal consciousness is that access consciousness is not *sufficient* for phenomenal consciousness. Block (1995) thinks it is not *necessary* either—phenomenal consciousness is fully independent from access consciousness. The significance of this claim is that phenomenal consciousness, the type of awareness associated with the hard problem of consciousness, would be fully dissociated from access consciousness, which is the notion of 'consciousness' used in global workspace theories. Suppose that access consciousness is best understood in terms of forms of attention (global, crossmodal attention and its subprocesses), and that phenomenal consciousness is fully independent from access consciousness, which is a plausible interpretation of these terms given standard definitions in the literature. Then Block's view would entail a serious degree of dissociation between attention and consciousness.

Block's (1995) distinction is controversial but it has played a central role in debates about consciousness. Block uses it to make theoretical points about the nature of the mind in general. For instance, Block (2003) suggests that physicalism and functionalism regarding consciousness may not be rival theories, but answers to different questions. Physicalism would address issues about the neural basis of conscious experience, while functionalism would attempt to answer questions about the purpose of those access-conscious neural representations that are available for thought, decision making, reporting, and action. This proposal resonates with the type of approach we promote. In particular, we claim that many theories that have been considered to be incompatible or rival views may be part of a larger theory of consciousness—one in which consciousness and attention are largely dissociated.

The importance of Block's distinction is that a natural way to interpret it is as a kind of dissociation between attention and consciousness. If one assumes that all functional aspects of consciousness that are understood as global access to information concern access consciousness, then this characterization seems to perfectly define crossmodal attention and attentional processes more generally. According to this interpretation, Block would argue for a significant degree of dissociation between consciousness, understood as 'phenomenal consciousness,' and attention, understood as globally available 'access consciousness.' This would not be a far-fetched interpretation for several reasons. Block claims, for instance, that there can be phenomenal consciousness without access consciousness, and vice versa, which obviously entails that they cannot be identical (see Block 1995, 2002, 2007). If access consciousness is interpreted in terms of attention, as we believe it should be, Block's proposal entails that attention cannot be identical with phenomenal consciousness.

The dissociation interpretation is also supported by how opponents of the distinction between phenomenal and access consciousness understand it. Dennett (2005) rejects the distinction between access and phenomenal consciousness because, according to him, there seem to be no good theoretical reasons to accept the mysterious notion of 'phenomenality.' More specifically, Dennett argues that consciousness must be defined functionally, and that the most plausible way to define it is as a form of global access that permits reportability and the use of information for several cognitive processes, including memory, thoughts, and action (for similar accounts, see Baars 1998, Nunn 2009). Dennett also argues that consciousness is a kind of functional monitoring that broadcasts information, and that it is intimately associated with language because this monitoring requires a common format for the manipulation of semantically structured contents, which is something language can do. Dennett's view favors a reductive-functional account of an identity thesis. Attention to globally accessible contents *is* consciousness. Whatever 'phenomenal consciousness' is, accordingly, it must be either reduced to global access or understood as a form of illusion.[4]

There are at least two possible ways to interpret this reductive view. One interpretation is that consciousness is identical to attention as a type of monitoring function. This view would deny that there is conscious attention—because consciousness is just attention with global access. An alternative view would hold that consciousness is identical with a very specific type of conscious attention: only the kind that can be readily available for linguistic formatting. The former view would attribute consciousness to creatures that are capable of attending in ways that are similar to the way humans do. The latter view would hold that the type of attention that is conscious might be a uniquely human competence that depends on language.

These implications are important when we begin to think about the evolution of consciousness, which we will explore in chapter 5. Because of Dennett's commitment to linguistic report as a mark of attention to contents that are globally accessible, his view can be considered a higher-order view—specifically, a higher-order *description* view. Higher-order theories are the topic of the next subsection, but worth noting now are the nontrivial constraints that identity views place on theories of consciousness; in particular, for example, the need for linguistic skills (see Dennett 1991). Other versions of the identity view ignore or deny access consciousness altogether and focus exclusively on phenomenal consciousness (see J. J. Prinz 2012). In general, identity views pay the price of significantly narrowing the scope of debates about consciousness, attention, and conscious attention—one of the reasons we favor dissociation. Obviously, this is not sufficient to constitute a decisive argument against identity views, and admittedly, an advantage of identity views is their parsimony. The case for dissociation must also be based on theoretical and empirical considerations. We believe that powerful empirical and theoretical considerations support a dissociation between consciousness and attention.

Given the many ways to interpret both 'consciousness' and 'attention,' how can it be possible to decide whether a specific finding supports their dissociation or not? The only solution to this problem is to create a more neutral framework to assess the evidence, relying on extant theories and theoretical findings. Here we focus on the current theoretical framework for definitions of 'consciousness.' Two theoretical issues deeply related to the dissociation between consciousness and attention are the unity of consciousness and the nature of global access. We discuss the unity of consciousness in section 3.4, because we think it is particularly important to give an account of the phenomenology of conscious awareness. With respect to criticisms of the distinction between phenomenal and access consciousness, the nature of global access seems to be a much more central issue. Block, for instance, does not make his distinction depend upon an account of the unity of consciousness. So it is to that issue that we now turn.

Block's main argument in defense of the distinction between phenomenal and access consciousness is the *overflow argument*, based on research concerning iconic memory (Block 2011).[5] The argument appeals to the findings by Sperling (1960), suggesting that the phenomenology of visual experience contains more information than that which can be perceptually accessed for cognitive processing. The central claim is that phenomenal consciousness is richer in content than access consciousness; more can be experienced than can be accessed for cognitive processing, including its reporting, at any given time. Critics of this proposal have pointed out that change-blindness experiments seem to contradict this hypothesis. Change-blindness experiments suggest two possibilities: either there is a process of specification (or precisification) from working memory to an iconic-*like* memory (to be distinguished from early visual iconic memory)—or a similar process of completion from fragmented representations (e.g., see Kouider et al. 2010 and references therein). Block's response to this criticism is that the type of dissociation he defends between access and phenomenal consciousness best accounts for evidence that supports overflow, rather than the hypothesis of a process in which contents become more precise. Regardless of who is right about this particular issue, there clearly is a role for conscious attention as either an interface between access and phenomenal consciousness or as a process that makes contents specific.

How should the nature of such a Blockean access be understood? First, according to Block, global access consciousness seems to be limited, in comparison to the amount of information that can enter phenomenal consciousness. But access consciousness cannot be a subpersonal and highly fragmented process, because this would mean isolated cases of access with no cognitive impact at the personal level are also cases of access consciousness. That would trivialize the notion, so a balance is required. While the contents of access consciousness must be integrated and semantically assessable for cognitive processing at the personal level, they need not be as rich as the contents of phenomenal consciousness.[6] This interpretation of the distinction between access and phenomenal consciousness suggests that Block is committed to three forms of attention: unconscious attention, access conscious attention, and phenomenal conscious attention.[7]

In sum, it is plausible to characterize Block's distinction between phenomenal and access consciousness under a Type-C dissociation account. If this interpretation is correct, he comes very close to endorsing a full dissociation account, because he argues that these forms of consciousness can be fully dissociated. Type-C dissociation views state that attention is not necessary for consciousness and that there are many types of conscious attention. It is likely that at least some form of crossmodal attention is necessary for consciousness, but we do think it is plausible to interpret Block's distinction in terms of a Type-C CAD.

Block says that access consciousness is limited to around four items, and that phenomenal consciousness—with its iconic memory—is richer and contains much more perceptual information than can be processed by attention. So there seem to be two types of *conscious attention*: phenomenal conscious attention and access conscious attention. Since, according to Block, phenomenal consciousness can occur without access consciousness and vice versa, selective attention may not be necessary for phenomenal consciousness, which makes Block's view a Type-C dissociation account.

Could phenomenal conscious attention include most forms of access conscious attention? Not under Block's account of the super-blindsighter example (Block 1997), which he uses to motivate the distinction between the two types of consciousness. A super-blindsighter has the capacity to process visual information in a manner that is functionally and behaviorally equivalent to someone with normal conscious vision, but without *any* phenomenal consciousness. These functionally equivalent visual cognitions are best understood in terms of nonphenomenal *access* conscious attention. Since a fully dissociated phenomenal conscious attention could be functionally equivalent to access conscious attention, this claim could also be interpreted as favoring full dissociation: conscious attention is not a natural kind, because there are two processes that differ radically in their properties but are nonetheless categorized under the same term. Therefore, the least dissociative interpretation of Block's proposal seems to be a Type-C dissociation.

An alternative way to disambiguate unconscious attention from conscious attention appeals exclusively to *representation*, rather than to how it interacts with the different forms of memory and the notion of phenomenal character, as Block's proposal does. The main idea behind such proposals is that there is a difference in representation between being conscious of a specific content—for example, that something is moving to my left and that it might be a bee—and being aware that one is having an experience with such content. In terms of representation, one might say that the experience's content is embedded in a representation that makes such content an experience that one is experiencing. Higher-order theories are all based on some formulation of this idea. We examine them in the next section, and explain why the best way to understand such theories is also in terms of dissociation.

3.1.2 First-Order and Higher-Order Theories of Consciousness

First-order and higher-order theories of consciousness can be classified in terms of supervenience and sufficiency claims (Byrne 2004). Supervenience claims concern the dependency relationships between different forms or levels of consciousness, and characterize the type of first-order or higher-order theory in terms of two views: 'representationism' and 'phenomenalism.' Sufficiency claims classify these views in terms of whether or not they count as higher-order theories. Supervenience and sufficiency claims, we shall argue, depend upon the dissociation between consciousness and attention. Alex Byrne (2004) argued that these debates were largely verbal disputes concerning ambiguous uses of the term 'consciousness' and 'what it is like.' Until now, no one, to the best of our knowledge, has interpreted this debate as a legitimate controversy concerning the degree of dissociation between consciousness and attention. We argue that this debate, like others in the literature, becomes much clearer and more useful once it is understood in terms of the spectrum of dissociation we propose.

Let us begin with the supervenience claims. Although these claims are not essential for classifying views as first-order or higher-order theories of consciousness, they do delineate the different commitments that first-order and higher-order theories have with respect to function and representational structure, associated with attention and what Chalmers (1996) calls the "easy problems" of consciousness. Representationist views, also known as *intentionalist* views, propose that the phenomenal or qualitative character of experiences entirely supervenes on their representational content: there can be no differences in phenomenal character without differences in representational content. Since representational content can be described functionally, as perceptual/cognitive functions from inputs to outputs, the representationist view about phenomenal content is more amenable to reductive views—those that characterize phenomenal content in terms of access consciousness, that is, access to functionally determined contents. For this reason, the identity view and the Type-A dissociation view are more easily characterized in terms of representationism. This does not, however, entail that the identity view can *only* be characterized in representationist terms; for example, J. J. Prinz (2012) avoids such characterizations, although that approach is not without problems either, as we shall argue throughout this chapter.

Phenomenalism (also known as *phenomenism*) is any view that denies the representational supervenience claim. Equivalently, phenomenalism denies that the representational supervenience claim holds necessarily for all experiences. Thus, there is room for degrees of phenomenalism, going from strong denials of representationism—the claim that representational content is irrelevant for phenomenal content—to weak denials. For instance, the claim that most phenomenal content supervenes on representational content, but there are exceptions,[8] would be a weak denial. The typical example in support of phenomenalism is the inverted spectrum (see Block 2003).

In the inverted qualia scenarios, of which inverted spectrum scenarios are one example, the qualitative character of two color experiences is inverted, even though the representational and functional aspects of the experiences seem to remain invariant. These include scenarios in which two subjects are having different experiences but sharing the same representational contents, or scenarios in which the same person undergoes some kind of inversion. In Block's inverted earth scenario, for example, the representational property associated with the linguistic content of 'red' remains invariant but involves a change between two experiences of color at different times, associated with phenomenal concepts. Alternatively, one can envision a scenario in which the experiences are qualitatively identical, yet the representational content differs. These are important distinctions, because it is plausible to characterize a blindsighter as having human-like access consciousness while having no phenomenal consciousness. Because of the previous arguments, this distinction would entail a full dissociation between attention and phenomenality.

The distinction that parses views into first-order and higher-order theories concerns sufficiency claims. According to first-order theories, higher-order representation concerning the subject's thoughts or other metaperceptual representations are *not necessary* for phenomenal consciousness. Any experience is a phenomenally conscious experience, and having an experience *suffices* for phenomenal consciousness. What it is like to see red is just having the experience of seeing red. Higher-order theories deny this sufficiency claim; having an experience is necessary but not sufficient for having phenomenal consciousness, that is, experiences with a distinct qualitative character. Higher-order theorists, therefore, have to postulate an extra cognitive ingredient to produce phenomenal character. This extra ingredient is either a *thought* that one is experiencing a phenomenal property, or a state with the disposition to produce such a thought (see Carruthers 2000, Rosenthal 1997), or a *perceptual-like* representation that one is having an experience (see Lycan 1996).

Higher-order theorists propose that having an experience *simpliciter* is very different from having an experience that is experienced as one's own. This higher-order representation is what makes an experience a phenomenally conscious experience: something it is *like* for an individual to have that experience. So, according to higher-order theories, having an experience does not suffice for being aware of it—a higher-order representation is necessary to make it conscious.

The supervenience and sufficiency claims yield different accounts of first-order and higher-order theories, each of which entail a different degree of dissociation between consciousness and attention. Representationist views can be understood functionally. A first-order representationist may characterize phenomenal content functionally, in terms of accuracy conditions. A higher-order representationist may characterize semantic content functionally, also in terms of accuracy conditions, although that account has to add the necessary condition that the first-order representation be embedded in another representation for it to have *phenomenal* content. Another, probably less well motivated, possibility would be a phenomenalist account of first-order experience that nonetheless requires higher cognition to produce fully conscious phenomenal content, thereby opening the possibility of unconscious phenomenal experiences—for example, experiences of red that are not conscious because they are unattended—an alternative that leaves open a vast amount of logical space for dissociations between consciousness and attention.

It is very plausible to characterize representationism in terms of access, which also yields different possible dissociations between consciousness and attention. In the first-order case, one has access to the contents represented in experiences by some sort of tracking function that detects these contents, or some function that accesses those contents by some representation-constituting relation. This representation-relation can be further constrained in terms of immediate acquaintance or causal rapport with stimuli. For a first-order representationist, being in such an acquaintance state, or having such causal rapport with stimuli, is sufficient for phenomenal consciousness. Although this seems to entail a very strong correlation between consciousness and attention, there is still the possibility of subpersonal forms of attention, or attention to contents that are not well integrated, which would not entail conscious awareness, thereby allowing for some forms of dissociation. In the higher-order case, by contrast, these functions need to be embedded in another representation: they need to be embedded in a representation that represents the experience *as an experience one is having*. So just having the first-order representation does not suffice for phenomenal consciousness, and all attentional processes of the first-level type are unconscious—entailing a much more severe form of dissociation in the higher-order case.

Now we are ready to put the spectrum of dissociation to work to shed some new light on this debate about consciousness. As we saw from the previous discussion, a question that is intrinsic to the first-order/higher-order debate is this: What is the status of experiences that are not conscious, according to the higher-order theorist, but nonetheless have representational content? About the supervenience claims, it is desirable to have a broad enough theory of consciousness to explain an array of semantic views. For instance, because of the importance of the inverted spectrum and similar cases for arguments in favor of phenomenalism, it seems useful to include different types of accuracy conditions. More specifically, the phenomenalist might want to explain what kind of content is represented in cases where there is access consciousness without phenomenal consciousness—assuming, that is, that this distinction is accepted.

Some philosophers (e.g., Brogaard 2011b, Chalmers 2003b, Siegel 2010) have proposed that there are some accuracy conditions that depend on the perspective of the conscious subject, while others may be independent of perspective. This distinction, though controversial, may help illustrate the difference between conscious and unconscious contents in a higher-order theory; it therefore seems pertinent to an analysis of the possible dissociation between consciousness and attention in the context of higher-order theories. One can think of these two kinds of accuracy conditions for conscious and unconscious content in terms of centered and uncentered properties: that is, properties that are centered on a world marked with an individual at a given time, versus properties at some possible world. For instance, Brogaard (2011b) argues that *being circular* could have two different accuracy conditions: the property of *being circular-shaped relative to a perceiver* (=centered), and the property of *being circular-shaped*, regardless of who is the perceiver (=uncentered). In the first case one could be viewing an ellipse from a certain angle that gives it the appearance of a circle, while in the second case the object is a circle regardless of whether anyone is perceiving it. According to this proposal, two observers may be looking at a 'red-w apple' (where 'red-w' is the wide content that would make their assertion 'the apple is red' veridical about the linguistically available content or possible-world content). The observers, however, can have different experiences of the color of the apple, say 'red-n(me)' and 'red-n(you)' (where 'red-n' is the narrow content of their assertion regarding the color of the apple, and concerns their experience of such color, a content centered on two different observers, e.g., 'me' and 'you').[9] In this way, one can explain why we can have the same accuracy conditions for wide content (e.g., 'red-w,' 'being circular-w'), and differ with respect to narrow content (e.g., 'red-n(me)', 'red-n(you)'). Cases of inverted spectrum are fully compatible with this relative-content semantics.

With respect to attention, it seems plausible to say that attention is always *attention to* contents—objects or events and their properties. If the debate between first-order theorists and higher-order theorists is reinterpreted in terms of the spectrum of dissociation, then what higher-order theorists argue is not as implausible as some have suggested.[10] The higher-order theorist seems to be saying simply that attending may cause experiences that are not conscious. Although the use of the term 'unconscious experience' is unfortunate, the higher-order proposal is not implausible once it is understood in terms of unconscious attention to contents. Attention is access to contents without the 'what is it like' character that defines phenomenal consciousness. If interpreted this way, attention to contents that are not phenomenally conscious would always be wide, while attention to phenomenally conscious contents would always be narrow. This form of dissociation in terms of content, however, seems too strong because the overlap between consciousness and attention may be more robustly integrated than the proposal suggests. Nonetheless, it is one way of making sense of the higher-order theories, and it entails dissociation between consciousness and attention.

Thus, one could argue in favor of a higher-order theory based on considerations about content. One option is to present such a theory as either a Type-A or Type-B CAD. If it entails a Type-A CAD, the theory implies that attention is necessary, but not sufficient, for phenomenal consciousness. If it is characterized as a Type-B CAD, it holds that attention is neither necessary nor sufficient for conscious awareness, and posits a single kind of conscious attention. In both cases there is only one type of conscious attention, because in all higher-order theories there is only one 'sense of what is it like to have an experience'—that is, one higher-order representation involving the first-person perspective.

Should higher-order theories be interpreted in this way? An advantage of stating such views in terms of dissociation is that it makes the contrast between first-order and higher-order representations more intuitive, explicitly connecting them to empirical findings on attention and consciousness in a much more straightforward way. However, to validate this approach we must scrutinize other aspects of the theory. For instance, the type of representation responsible for phenomenal consciousness would have to be a perspective-based type of representation, and it is not entirely clear how to model these representations without appealing to some primitive notion of phenomenality (such as first-order theorists claim) or to a first-order representation of phenomenal content in an immediately perspective-based way. This introduces the problem of subjectivity and introspection into our study of consciousness, which exacerbates the dissociation between consciousness and attention and further complicates their relation, as mentioned in chapter 1.

One could object to the classification of higher-order theories in terms of dissociation by claiming that this way of characterizing contents depends wholly on the distinction between access and phenomenal consciousness. In response one may say that although there is an obvious connection with the distinction between access and phenomenal consciousness, because of the emphasis on consciousness and attention, there are crucial differences concerning the representation of contents in higher-order theories, which entail dissociation and do not depend on this distinction. For example, the centering of contents in first and higher-order theories is entirely different. In the first-order case, it depends either on some kind of primitive property of the experience, or on a primitive relation that makes the content accessible to an individual from her perspective. In contrast, all higher-order theories dispense with this primitively understood component, requiring metarepresentations of some form. Moreover, an unconscious representation, for the first-order theorist, is not an experience and is not even accessible for semantic evaluation. The higher-order theory, however, is more flexible in this regard and allows for the distinction between unconscious attention and conscious attention to the same content, or the same experience.[11]

It must be emphasized that higher-order theories do not postulate a representation of a self in order to explain the metarepresentation requirement for phenomenal content.[12] Self-consciousness seems to require more than just metarepresentation or higher-order thought—an issue we address more thoroughly in section 3.4. So there is considerable room for introducing even more forms of dissociation, like self-conscious attention or self-consciousness without attention. Our main contention is that attention to contents without phenomenal consciousness may be understood in terms of at least one type of dissociation. In fact, dissociation provides a very plausible way to construe higher-order theories. This claim may be reinforced by empirical findings concerning motor control and action selection, which will be explained further when we discuss agency in the context of possible dissociations between consciousness and attention.

Interestingly, *representationist* higher-order theories require less dissociation (i.e., Types A or B CAD) than first-order phenomenalist theories (i.e., Type-C CAD). Another result of the previous arguments is that higher-order *phenomenalist* theories require no more dissociation than first-order phenomenalist ones. This is counterintuitive, because a general complaint about higher-order theories is that they make distinctions where there are none, such as distinctions between perceptual unconscious belief and higher-order perceptual belief, or between unconscious and conscious experience. Thus, from a purely theoretical perspective, interpreting higher-order theories in terms of the spectrum of dissociation that we propose highlights aspects of extant theories that are otherwise difficult to appreciate. Higher-order theories of consciousness are really theories about *how* information is accessed: their main motivation is that conscious information is not accessed at the level of *immediate* representation, but rather at the level of *cognition* for report and action (Dennett 1991). Thus the distinction between the personal and subpersonal levels of information becomes important. The information that is being monitored or targeted by higher representations must have *some* degree of unity and integration; in principle, it must be in a format that makes it available for report and action, such as language or a language of thought.

According to higher-order theories, then, there is a mapping from a set of first-order representations to higher thoughts or experiences, both with their own degree of informational cohesion. In general, the contents of access consciousness require cognitive integration (e.g., mappings from allocentric to egocentric frames of reference, and modal and crossmodal solutions to binding problems). Moreover, all theories of consciousness need to account for the highly unified, stable, and integrated contents of phenomenal consciousness. We address this topic in the next section. But some final remarks are required to clarify how integration plays a crucial role in the dissociation between attention and consciousness for higher-order theories.

The overload objection against *actualist* higher-order theories—that is, theories which require the tokening or specific representation of higher-order thoughts in order to target a set of unconscious beliefs or contents at a time—is that it would take an implausibly large amount of higher-order thoughts to produce the rich phenomenology of conscious awareness. A way to circumvent this difficulty is by appealing to dispositions: experiences that are conscious are those disposed to produce higher-order representations of the sort that are available for higher-order thought.[13] A crucial question with respect to this issue is how scattered or integrated the targets of higher-order representations are. Presumably one can attend to and bind these contents unconsciously (e.g., via object files), as the empirical literature shows. Even if there are unconscious inferential relations that specify these contents, higher-order thoughts seem to be produced, or at least become available, by means of a noninferential process.

Unconscious, cognitively integrated doxastic attitudes (i.e., attitudes that are belief-like) that are used in action and motor control, therefore, are best understood in terms of *first-order attention*. So the question is, does the mapping between higher-order and first-order representations depend on a highly integrated set of such unconscious doxastic states? It seems plausible to say that states which are either targeted by higher-order thoughts or disposed to produce higher-order thoughts must be *globally accessible* (accessible to access consciousness) so that they can produce phenomenal consciousness. Alternatively, if the notion of phenomenal consciousness is dismissed, then the unconscious contents must be at least cognitively integrated in order to be access consciousness accessible. In either case, unconscious first-order attention must depend on a high degree of cognitive integration, according to higher-order theories of consciousness.

3.2 Cognitive Integration and Experience: Subsumption and the Unity of Consciousness

Consciousness is frequently defined in terms of global access to highly integrated information. In this section, we examine that definition of consciousness in light of our proposed spectrum of dissociation. We argue that although consciousness is indisputably associated with access and cognitive integration, the unity of phenomenal consciousness and the global character of access to information are cognitively independent of one another. We explain why cognitive integration should not be identified with access to consciously accessible content, and also why the unity of phenomenal consciousness is distinct from access consciousness. The conclusion of this section is that these distinctions entail a severe form of dissociation between consciousness and attention.

3.2.1 Cognitive Integration and Access

According to Tyler Burge (2010), any legitimate mental representation must attribute properties to particulars at the organism level. Otherwise, the information is transmitted mechanically and subpersonally without truly representing the environment. Mental representations must satisfy other constraints in addition to this somewhat controversial one. For example, they must allow for cases of misrepresentation, and must be susceptible to combination with other representations. The contents of representations must be the propositional basis for psychological attitudes that structure behavior. They must also support counterfactual reasoning, which need not be conscious reasoning, thereby informing the cognitive agent's decision-making capacities. Clearly, these constraints on representations are fundamental for any theory of the mind, independently of considerations concerning consciousness and phenomenology. The intentionality of representations, therefore, seems to require a high degree of cognitive integration as well as access to complex and unified contents. But what is the relation between access and integration?

Beliefs, even unconscious ones, are doxastic endorsements of contents—a kind of epistemic assertion. Even if one repudiates the idea of unconscious belief, it is clear that a lot of our behavior, as well as the behavior of infants and animals, is guided by doxastic attitudes toward contents with accuracy conditions. A vast amount of successful behavior is guided by unconscious representations for motor control, an issue we expand upon in section 3.4. Access to information, therefore, is fundamental not only to explain the structure of mental states and their content, but also to provide a framework for basic epistemic achievements such as those involved in navigation.

However, information can be accessed in different ways. For instance, one can imagine a massively modular cognitive agent that is extremely successful at a vast array of behaviors, but is incapable of cognitively integrating the contents of the representations involved. One could say, because of the personal-level constraint, that these are not genuine representations. Alternatively, and perhaps less controversially, one could seriously question that such a cognitive agent has the capacity to form beliefs in a systematic way by deducing conclusions from premises or by inferring consequences. But in order to achieve sufficient integration and guarantee that most epistemic processes are not modular and isolated, all that must be accounted for is a measure of informational complexity among distributed areas that contribute to the production of unified representational contents. This type of integration may occur during navigation, and it is at least conceivable that in many organisms this happens without high-quality epistemic access to contents (e.g., the quality of access required in order to form beliefs).

So besides cognitive integration, global access imposes further constraints on mental content, particularly with respect to report, action, and memory. More specifically, the quality and degree of access to content for report, action, and memory, independently of its phenomenal character, is decisive in characterizing access consciousness. Take two behaviorally equivalent agents. One is massively modular, and through independently performed computations and a complex biology, it succeeds in many tasks. The other system performs the same functions but does so using a cognitively integrated and "centralized" set of representations. Suppose that these highly integrated representations depend upon systemwide mappings, as do those in the human brain that facilitate motor control. In the second agent, such a high degree of integration could be used to explain the functions that lead to successful behavior in terms of composite representations, rather than strictly mechanical or highly independent systems.

Without knowing whether this latter cognitive agent has access to such information, however, one cannot describe such information processing as access consciousness. In particular, epistemic access is global in a more robust sense than the notion of 'centralized representation' can capture, because global access requires capacities to report, act, or memorize those accessed pieces of information; and these are options that a common format for centralized representations cannot produce on its own. It seems, then, that even if one grants that the integrated agent represents the environment, it is still not clear that such an agent has access for thought, report, action, and memory to the contents that are being represented.

How should one understand access in relation to degree and quality? Examples from the literature help answer this question. Since access is understood as availability for thought, report, action, and memory, one could rank degrees of access for each of these capacities assuming equally integrated contents across rankings. For example, blindsight patients can attend and respond to features, even though they have no phenomenal experience of these features. One can imagine, then, three distinct cases of blindsight, all with the same high degree of cognitive integration: (a) a blindsighter, Bob, who is reliably good at succeeding at tasks, but has no access whatsoever to the information involved in performing these tasks; (b) a super-blindsighter, Lucy, who is slightly better than Bob in performing tasks and has moderate access to information (she can tell there is an object with a certain orientation, for instance, but not know the identity of the object or its precise orientation); and (c) a super-duper-blindsighter, Maria, who has incredibly detailed access to information, as good as that of standard human beings—but without having any experiences with phenomenal character.[14]

One can think of more complicated cases like partial or modally specific zombies, with the degrees of access just mentioned. The main point is that Bob, Lucy, and Maria may all be very good at hitting baseballs, but only Maria can have beliefs about hitting baseballs in specific orientations. Bob may have no idea what he is hitting, while Lucy may have a very poor description of the object. Because of these examples, one can use degrees of access to distinguish cognitive integration from access consciousness. Just as there are degrees of cognitive integration, from modularly independent information to highly unified representations, access also comes in degrees. Crucially, the degrees of access seem to depend on the degrees of *specificity* of available semantic information.

If one identifies the degree of specificity of access consciousness with the richness in content of phenomenal consciousness, then typical examples concerning zombies would show that unconscious epistemic agents may possibly have the same degree of accuracy and richness in semantic information. Actually, a super-Maria would know with a lot more precision than the standard human the exact shade of an object's color, even though she cannot have any experiences associated with color. Assuming that Maria apprehends the same degree of precision as normal human subjects, one could say that her access consciousness is functionally identical to our phenomenal consciousness—although, by assumption, she has no phenomenal experiences. In the case of super-Maria, such access would be more precise than the access provided by the conscious experiences of humans. For normal subjects, rather than imaginary ones, it could be that access consciousness, compared to phenomenal consciousness, is less precise and not as rich in information. (This is what Block, relying on the empirical considerations mentioned previously regarding memory capacities, might claim.) Regardless of whether this empirical claim is true or false, it is clear that one can distinguish the extent of cognitive integration—the degree to which different representations are integrated to produce new ones—from the degrees of access consciousness.

This is an important distinction in general, but it is particularly relevant with respect to recent attempts to explain consciousness in terms of the degree of cognitive integration. Consider, for example, Tononi's (2008) *integrated information theory* (IIT), which is endorsed by Koch (2012a). On the basis of the previous discussion, it seems clear that defining 'access consciousness' or 'phenomenal consciousness' exclusively in terms of cognitive integration, as IIT proposes, would misconstrue what these terms mean. Access consciousness is the availability of *semantic content*, that is, propositional content capable of anchoring attention. This is why the degree of specificity between cognitive integration and semantic unity differs. The main implication of the previous discussion seems to be that only epistemic agents can have access consciousness. Access consciousness is a necessary condition for epistemic agency; phenomenal consciousness matters for epistemic agency only insofar as it affords access. Cognitive integration is necessary for adequate access consciousness, but it is *not sufficient*. A highly integrated "cognitive" system, like the Internet, presumably lacks access consciousness for this reason, and therefore fails to qualify as a cognitive *agent*.

With respect to the dissociation between consciousness and attention, these considerations reinforce our previous claim that access consciousness is associated with most forms of attention, which means that many forms of attention could occur in the absence of phenomenal experiences. This may be a controversial claim for those who identify consciousness with attention, but as we argue, there are powerful theoretical reasons to accept it; most accounts of how to define 'consciousness' entail this claim in one way or another. We would add to this familiar distinction that agency is a distinctive aspect of access, but not of cognitive integration. Just as it is odd (and inadequate) to say that the Internet could attend to the statistics concerning energy consumption, it is odd (and inadequate) to say that Maria is *not* attending to the baseball as she prepares to hit it, even though she lacks phenomenal consciousness. Attention, moreover, seems to anchor demonstrative thoughts. Maria, being a super-duper-blindsighter, can attend to the baseball *as such*, unlike Lucy, who can attend to the baseball only in a very coarse way, and cannot attend to it *as a baseball* coming in her direction with a particular spatiotemporal trajectory.[15]

Maria can not only attend to features of objects and identify such objects, she can also respond to questions about them and report the information she has available. She can have attitudes toward the contents she is accessing. Critically, Maria is capable of endorsing certain contents as propositions she believes, and is responsible for her epistemic assertions with respect to other epistemic agents. How much self-awareness and self-knowledge this type of epistemic agency requires is a question we analyze in section 3.4. What we want to emphasize now is that as long as Maria has a very rich access consciousness, similar or superior to that of the standard human being, she could have a very sophisticated kind of epistemic agency, even in the absence of phenomenal consciousness.

A brief clarification: we are not claiming that modularity is incompatible with access consciousness. High degrees of cognitive integration and global access do not entail either holistic networks or massive modularity. On the contrary, it seems plausible that early stages of information processing are computed in isolation by different specialized modules, and then integrated for, among other purposes, navigation and motor control. Likewise, many modular computations occurring at a given time are compatible with a broadcasting system that globally accesses information based on the previous integration of such information for semantic and epistemic purposes. Although this broadcasting *requires* previous cognitive integration, mere integration does not suffice.

3.2.2 Access Consciousness and the Unity of Consciousness

Suppose that access conscious attention is global access with at least a moderate level of detail, such as Lucy's. Then there could be two basic kinds of unconscious attention: non–globally accessible attention and very poorly detailed attention. Should these forms of attention be called 'attention' or should they be called something else, like 'orientation'?[16] This returns us to our earlier considerations concerning representation at the level of the organism: accuracy conditions and integration. It seems that as long as these forms of attention function in the way that higher forms of attention operate—discriminating, selecting, and inhibiting stimuli—they all deserve the name 'attention.' The important difference is that they are all unconscious and epistemically degraded forms of attention.

So it seems that only global access that is at least moderately detailed deserves the name 'access conscious attention.' Is the type of integration and epistemic access required for access consciousness sufficient for the unity of phenomenal consciousness? Functionalists would find this question senseless. Whatever degree of detail and epistemic integration access consciousness has is *exactly* the degree of detail and integration that phenomenal consciousness has, because phenomenal consciousness is to be understood functionally. But those who accept the distinction between access and phenomenal consciousness will not accept such a reductive account.

We analyzed the distinction between access and phenomenal consciousness above, arguing that it entails a severe dissociation between consciousness and attention. We then explained the type and quality of integration required for access conscious attention. Here we focus on *phenomenal conscious attention*. If one accepts the distinction between access and phenomenal consciousness, how should quality and degree of integration be distinguished? One option is to make the distinction by appealing only to detail and richness—that is, epistemically available semantic content richness. According to this proposal, the detail and richness of access consciousness is moderate, while the detail and richness of phenomenal consciousness is very high. This is the conclusion of Block's overflow argument.

Examples that illustrate why demonstrative thought and reference may be constitutive of the phenomenology of conscious attention introduce problems that concern specifically epistemic issues and generate difficulties concerning cognitive function. In John Campbell's 'sea of faces' example (2002), one consciously sees, or at least has a phenomenal visual experience of, a collection of indistinct faces. Assuming, for the sake of the example, that visual access consciousness is a lot more detailed than phenomenal consciousness, even if one had epistemic conscious access (in terms of access consciousness) to the distinctiveness of these faces, one would not be able to think demonstratively of a particular person (e.g., *that* woman) until one had phenomenal conscious attention access to her face. So phenomenal conscious attention seems to be necessary for demonstrative thought.

This is a controversial way of interpreting Campbell's example, but it is a useful way to think about the dissociation between epistemic agency and phenomenal character. With respect to this way of understanding the 'sea of faces' example, it seems that at least two senses of 'attention' must be distinguished, as we suggested above, with clear implications for our proposal concerning dissociation. The distinction, in this case, would be based not on access to semantic information without phenomenality, but rather upon the constraints on *demonstrative thought* that seem to require phenomenality. The idea is that access consciousness, even if highly detailed, could operate descriptively and functionally and with great precision, but it would be incapable of supporting demonstrative reference. One type of attention could be called 'functional attention'—which subjects with blindsight have, because it is defined strictly in terms of selection, inhibition, and targeting— while the other one, 'phenomenal attention,' affords a distinctive 'what it is like' character to attended contents and establishes a *direct acquaintance relation* with such contents.

Campbell uses this example to show that, since the main purpose of conscious attention is to target a specific content demonstratively, one needs a relational account of attention rather than a strictly representational one. Declan Smithies (2011) challenges Campbell's (2002) relational view of attention for demonstrative thought on the basis of a distinction between functional and phenomenal attention. Smithies proposes an epistemic interpretation of phenomenal attention, which requires availability for rational inference, particularly concerning immediately justified belief. Although we agree that epistemic agency is crucial for the adequate characterization of this case, we disagree with Smithies's proposal for explaining such a distinction for the following reasons.

Smithies says that the difference between his and Campbell's accounts of conscious attention is that they operate at different levels of psychological explanation: Campbell's account of attention, according to Smithies, is computational, whereas Smithies's account is rational (Smithies 2011, 19). Smithies refers to Dennett's (1969) distinction between the subpersonal (or strictly computational) level and the personal (or rational) level to clarify what he means by this distinction. But for reasons stated above, this is not a helpful clarification. In particular, the personal and subpersonal levels could be understood in terms of cognitive integration, which does not suffice for access to contents. If the distinction concerns epistemic normativity (i.e., contents that *should* be epistemically endorsed) and agency, such that the computational account is not epistemically normative while the rational is, then one would have to argue that cases of super-blindsighters operate at a level of psychological explanation that is strictly computational. But as we argued, this is implausible, because Maria seems to be a responsible epistemic agent. So phenomenality may not be necessary for rational capacities and for being a responsible epistemic agent, suggesting that Smithies's distinction may be orthogonal to the distinction between conscious and unconscious attention. In any case, this distinction also suggests that there must be a form of dissociation between computational and rational attention. Whether or not phenomenal conscious attention is necessarily rational remains a more controversial issue. For instance, pain experiences or conscious visual experiences of color may not necessitate the deployment of inferential capacities.

If phenomenality were necessary for responsible epistemic agency, then one would not be able to make sense of epistemic agency and normativity without phenomenal conscious attention. We shall argue that this is clearly not the case. Although phenomenal consciousness may very well provide the richest kind of normativity, the kind necessary for moral responsibility, access consciousness *suffices* for epistemic normativity. A super-blindsighter with a rich enough access to contents has all she needs to be a responsible epistemic agent. A detailed argument for this claim is provided in section 3.4.

We are not claiming that an epistemically rich and detailed access consciousness is sufficient to eliminate the theoretical need for phenomenal consciousness. The overflow argument could be interpreted as implying that having very rich access consciousness would make it impossible to distinguish between phenomenal consciousness and access consciousness. But even if this were true—for the practical reason that reportability seems to be crucial to identify phenomenal consciousness, and reports require access—the distinction between access and phenomenal consciousness, as well as the dissociation between consciousness and attention, can still be decisively identified based on other, nonepistemic considerations.

If epistemic normativity and rich access are not uniquely distinct features of phenomenal consciousness, then what is its distinguishing feature? Notice that even if epistemic normativity and rich access were the most distinctive features of phenomenal consciousness, we would still have dissociations between consciousness and attention, which itself is our main point in this chapter and throughout the book. There would be, for instance, subpersonal attention, access conscious attention, and phenomenal conscious attention. There is an argument in the literature on consciousness, however, suggesting that what is distinctive about phenomenal consciousness is neither rationality nor rich epistemic access, but the difference in the relational *structure* that unifies contents—that is, that the difference between global epistemic access and phenomenal unity is in the way their contents are unified. We shall argue that making the distinction between access and phenomenal consciousness along these lines could lead to a much stronger type of dissociation between consciousness and attention—either very severe (Type-C) dissociation or even full dissociation.

Unlike the examples concerning demonstrative thought, the question concerning the structure of access and phenomenal consciousness requires an assessment of how contents and experiences are integrated or unified at a given moment in time. The pertinent considerations regarding the unity of consciousness and the structure of global access are independent of those concerning the specificity or epistemic status of demonstrative thought. We argued that in the case of access consciousness, because of its association with verbal report, a high degree of cognitive integration plus semantic structure (perhaps characterized in terms of compositionality or even inferential role) seems fundamental—at least in the case of rich epistemic access. According to some authors, however, the situation is radically different with respect to phenomenal consciousness.

The unity of phenomenal consciousness should define what it is like to have a set of experiences at a given time. As Block (1995, 230) says, "the totality of experiential properties of a state are 'what is it like' to have it." Inferential structure and semantic compositionality, and even conceptual content, seem to be too strong and generally inadequate to characterize the relation responsible for *phenomenal* unification. One does not experience the fragrant smell of a flower and its beautiful colors as components of a set of inferences or semantic categories concerning fragrance and flowers. Franz Brentano, for instance, thought that besides intentionality, the mark of the mental conscious experience brings with it as well a robust and unique form of unity, such that anything that is experienced by a subject seems to be part of an overall experience.[17] The question is, what relation unifies conscious experiences from very different sources (sensorial, emotional, mnemonic, etc.) in such a robust way?

Tim Bayne and David Chalmers (2003) explore several options, and propose a relation they call 'subsumption' defined in largely negative ways. Subsumption, according to Bayne and Chalmers, is the relation that unifies any set of conscious experiences that a subject has at a time into a cohesive overall experience, and provides a kind of unity that *cannot* be explained exclusively in terms of neurological, spatiotemporal, object-based, functional-representational, self-based, or rational relations.[18] Cognitive integration of a very high degree—which would include all the relations just mentioned—is *insufficient* for phenomenal unity. Access consciousness is *also* insufficient to produce phenomenal unity, for the following reason. Access consciousness seems to be *conjunctive*, according to Bayne and Chalmers. One accesses a *conjunction* of contents, and those that are access unconscious are not part of the access set. Even if one includes contents that are *accessible*, rather than accessed, the relation that unifies them is conjunction; one can access the identity of an object and its location and other contents that are accessible, such as its shape.

Bayne and Chalmers say that the difference between subsumption and access *entails* the distinction between phenomenal and access consciousness. If there is access conscious attention and phenomenal conscious attention, this means that subsumption entails that identity theories of attention and consciousness are ambiguous and ultimately false. Subsumption is a lot more like part-hood than conjunction. Necessarily, when one has a phenomenally conscious experience, it becomes part of an overall phenomenally conscious experience. So subsumption entails that access consciousness is insufficient to *unify* phenomenal experiences, and this in turn entails a form of dissociation between how contents are attended in access consciousness and how they are attended in phenomenal consciousness. Subsumption also seems to be a primitive relation, in the sense that it cannot be reduced to other relations and it holds necessarily for any conscious experience. If one experiences the smell of a flower and its colors, then there is a single phenomenal experience that subsumes them and determines what it is like to smell and see the flower.[19]

What kind of dissociation does the distinction between access conscious unity and subsumption entail? If Bayne and Chalmers are right, then one cannot give an account of unity if one favors a view that identifies all forms of consciousness with all forms of crossmodal attention. According to that view, consciousness would be just global attention. As mentioned, subsumption entails the distinction between phenomenal and access consciousness, and as we argued, this distinction entails a Type-C dissociation. Subsumption entails at least this degree of dissociation, but it could also entail the most severe one, full dissociation with no conscious attention.

Access attention is *functional* and conjunctive, but subsumption is neither functional nor conjunctive. If subsumption is the relation responsible for the unity of phenomenal consciousness, then there are two fully dissociated types of conscious attention. *Conjunctive attention* would be the typical kind of attention studied in psychology, which is defined functionally. The contents of conjunctive attention are targeted or selected from a set, and one can select one or another without their becoming part of a more integrated content. *Subsumptive attention* is uniquely distinctive of phenomenal consciousness. The contents of subsumptive attention are such that, necessarily, when one attends to one of them, one attends not to a member of a set, but rather to a part of a unified, total experience with a distinct phenomenal character inseparable from any of its parts. Thus, interpreted this way, subsumption seems to entail a full dissociation between 'consciousness' (defined in terms of subsumptive attention) and 'attention' (defined in terms of conjunctive attention).

This structural dissociation that emphasizes subsumption and conjunction may be interpreted in at least two alternative ways, both of which entail dissociation between consciousness and attention. One of them would characterize the structure of conscious contents not as happening one state at a time, but as being integrally subsumed by a 'field of consciousness.' According to this view, the conscious field is responsible for integrating and making accessible all conscious contents (Bayne 2007, 2010). The phenomenal field is, therefore, a way to explain why all phenomenal contents are integrated as parts of total experiences, because the field imposes such part-whole relations.

This way of construing field views entails a dissociation because attention would operate conjunctively and discretely either on conscious contents integrated in the field or on unconscious contents, as in cases of vision without awareness. As mentioned, attention operates not only conjunctively but also in a limited itemized fashion (Pylyshyn 1989, 2003b, 2007). Conscious attention, therefore, would be the parallel operation of the mechanisms of different types of attention on the conscious field. Since consciousness and attention would be dissociated according to this version of the field view, it is compatible with CAD; in fact, it entails a strong CAD.

A further complication that would make the dissociation more severe concerns language. If demonstrative reference is fundamental for the operations of attention, and language is fundamental to produce demonstrative thoughts (a controversial but by no means unpopular assumption), then the language capacity would be necessary for the interface of the mechanisms of attention within the conscious field. Such a conclusion, however, would entail that most species are incapable of having conscious forms of attention. It would also entail that conscious attention emerged as a consequence of the language capacity, an issue that we explore in chapter 5. For similar reasons, as we explained in chapter 1, invoking a primitive first-person point of view or a primitive form of subjectivity as the defining characteristic of conscious contents also entails dissociation, with considerations about self-awareness concepts and linguistic capacities for self-reference further complicating the picture.[20]

3.2.3 Types of Access and Unity in Terms of Consciousness, Attention, and Conscious Attention

It seems that if one defines access in terms of a functional selection of mental contents, then one must allow for subpersonal and unconscious forms of attention, modular-early stages, as well as cognitively integrated ones. This is by no means uncontroversial, but it is also a typical way of defining 'early vision' forms of attention. If by access one means semantically and perhaps epistemically normative contents available for thought, verbal report, and action, then it seems one must restrict the use of the term 'attention' to access consciousness as the latter is typically defined. These would be two different types of non–phenomenally conscious attention: unconscious attention and access conscious attention. One may have a strong philosophical aversion toward calling unconscious attention 'attention.' But a very important reason to call it 'attention,' and the reason why psychologists used the term in order to characterize many activities that occur at the unconscious subpersonal level, is that causally driven *selection* of features that become integrated with other features to support perception, memory, and action are inherently attentive processes.

Although access consciousness may not be sufficient for phenomenal consciousness, it seems to be sufficient for epistemic normativity. However, the philosophical resistance against unconscious forms of attention and non–phenomenally conscious forms of attention may depend on less normative considerations, which deserve some comment and elaboration. We will focus on Jesse Prinz's (2012) identity proposal because it rejects the 'phenomenal versus access consciousness' distinction without endorsing high-level cognition. Based on desiderata that include rejecting high-order interpretation, higher-order theories, dualism, and self-consciousness or self-awareness views, Prinz argues that consciousness should be identified with attention. Given that most extant accounts entail some form of dissociation between consciousness and attention, and in some cases quite a severe dissociation, as the arguments we offered so far seem to show, this is a problematic thesis. By 'attention,' however, Prinz means something like 'attention to information that is available for working memory.' Prinz denies that 'access *consciousness*' could be defined as attention to information available for working memory. This means that consciousness just is, for Prinz, information that is available for working memory, plus phenomenality. He writes: "I don't believe there is any form of access that deserves to be called consciousness without phenomenality. After all, access is cheap. When an ordinary desktop computer calls up information from a hard drive or responds to inputs from a user, it is accessing information, but there is little temptation to say that the computer is conscious. Information access seems conscious in the human case when and only when it is accompanied by phenomenal experience" (J. J. Prinz 2012, 5–6).

The blindsighter, according to Prinz, has phenomenal consciousness of inner speech, and this is how she can access information with phenomenality—or at least, *some* limited phenomenality; the blindsighter cannot have visual phenomenality. But what if the blindsighter has no phenomenal experience at all, and yet can access the information successfully to adequately represent her environment? Why deny that she is attending to the features she is *representing*, unlike the computer, which is not representing anything? (All the manifestations of a computer's capacity to "represent" information derive from our intentional interactions with such representations.) More important, Prinz uses a definition of 'access' that is implausible because it confounds representational-content access, as we defined it above, with simple *cognitively integrated* access, such as the kind we associate with the Internet or an ordinary computer. Therefore it does not follow that, because computers are not conscious, the blindsighter is not conscious, in some relevant sense of the word 'conscious.' The computer accesses information only *relative* to our interpretation of such information, whereas the blindsighter accesses information *based on her own capacities,* not relative to anyone else's interpretation. Crucially, the distinction between access conscious attention and phenomenal conscious attention is not eliminated by considerations concerning information integration, for the reasons explained in section 3.2.1, which clearly do apply to computers.

What would be an example of phenomenality without access? Block (1995) argues for what we consider to be a high degree of dissociation between access and phenomenal consciousness. He says that when one has a temporally lasting experience, such as awareness of a constant background noise, one may not be accessing the information while one is phenomenally conscious of the experience; and one also could access information without having an experience, as in blindsight. Based on the previous discussion on subsumption, could this distinction be interpreted in terms of access conscious and phenomenal conscious attention? In the case of the background sound, one lacks access conscious attention to the sound, but one has a phenomenal experience of it, according to Block, in the sense that one is distinctively experiencing it. In the case of the blindsighter, there is access conscious attention without phenomenality. One could argue that this means that there are two types of conscious attention (i.e., phenomenal and access), which is admittedly a quite controversial thesis. At the very minimum, it does seem reasonable to say that there is a dissociation between robust conscious attention, which necessarily involves access and phenomenality, as Prinz requires, and access conscious attention, which may not require phenomenality or subsumption. This would entail a Type-C CAD. If there is only access conscious attention, and phenomenality can be fully dissociated from it, then attention is always functional and unconscious, without phenomenality. Whatever phenomenality contributes cannot explain how contents are functionally accessed. (This position is compatible with forms of phenomenal consciousness eliminativism, as in illusory accounts of phenomenality, such as Humphrey 2011.)

These considerations suggest that a full dissociation between access and phenomenal conscious attention is at the very least a plausible theoretical option. In any case, most extant accounts entail at least some form of dissociation. A Type-C CAD would make room for further dissociations, such as between unconscious and conscious attention. There is also an intriguing possibility in the vicinity of these possible dissociations. If attention is deeply related to availability in working memory, which can only be defined conjunctively as access to information, according to Bayne and Chalmers (2003), then that means attention is a conjunctive, discrete, and additive phenomenon. The more contents that one attends to, the larger the conjunction of items in the access set and the more options for selection and retrieval.

By contrast, subsumption is not additive, and the purpose of subsumption is not discrete *selection*. Rather, subsumption accounts for the unified global phenomenality of the conscious stream. The more experiences that are subsumed, the richer the overall experience. Selection, inhibition, and the complexity of access seem irrelevant to the subsumption function. Since they seem to fulfill two entirely different functions, one way of interpreting Bayne and Chalmers's account of subsumption is that it entails full dissociation. Although there is the appearance of an overlap between phenomenality and access, there is no real overlap. Phenomenal conscious attention is not a natural kind.

Before we address some of the findings that support dissociation (besides blindsight, which has been discussed extensively in the literature), we should briefly focus on a question that is central to understanding phenomenal attention isolated from access—a possibility that is compatible only with Type-C CAD and full dissociation. The question is: what could be the cognitive role of having an experience without access consciousness? Subsumption allows for the possibility of having phenomenality without access. So, what would phenomenality without access be like, and what could be its function? The question, put this way, is already tendentious, since it is controversial to say that phenomenality has a specific function. But we can weaken the latter claim by making it a loosely defined role, the 'phenomenality subsumptive' role.

My experience of having intense pain plays a role that is independent of the access conscious functions such experience seems to play in most circumstances, such as verbally reporting that I am in pain, voluntarily identifying the area that is damaged, or inferring what could have caused the particular intensity of the pain. Suppose there is a creature that lacks any form of cognitive integration. This would preclude the possibility of access consciousness, since that requires sophisticated cognitive integration and global broadcast. Furthermore, suppose that this creature is capable of experiencing pain, of the same quality and degree as humans, but that pain is the *only* experience it can have. One might call such a creature a 'pain blob.' The most minimal description one can give of these experiences is in terms of a scale, from 'good' (no pain) to 'very bad' (intense pain). This is not the way such a creature would make sense of its pain, but it is a plausible way of approximating how such a creature would rank pain. Let us call this a strictly phenomenal quasi-normative role.

Now imagine a human being that is capable of performing and representing the functional aspects of pain (she knows what pain behavior means, knows how to mimic it, and knows that it relates to bodily damage) but lacks *any* experience of pain, in spite of the fact that she is capable of experiencing *everything else* a human is capable of experiencing.[21] Her situation seems to be one of access to information with accuracy conditions that can be semantically evaluated (concerning pain behavior), without any internal connection to how pain feels. More specifically, her situation is one in which the experience of pain cannot *guide* her responses *automatically, effortlessly, and implicitly*. This normative feature of phenomenality seems to be a lot more important than mere access to contents. It seems that the difference between the 'pain blob' and the painless human is one of experience without access versus access without experience. We explore this further in the concluding sections of this chapter. For now, we reiterate that the main result of our arguments is that, given the current state of the debate on consciousness and attention, a severe form of dissociation between them is very likely.

3.3 Consciousness without Attention: Theoretical and Neuroscientific Considerations

We have offered three theoretical reasons in favor of the dissociation between consciousness and attention. The first is based on Block's distinction between phenomenal and access consciousness. The second is based on interpretations of higher-order theories. The third is based on considerations concerning the sufficiency of access consciousness for epistemic normativity and the relational structure of phenomenal unity. In the previous chapter we presented empirical support for several distinctions between types of attention. In this section, we briefly cover empirical evidence in favor of the kind of dissociation for which we have been arguing. Our own empirical argument based on evolution and findings across species is presented in chapter 5, but the following considerations show that the evidence in neuroscience and cognitive psychology also strongly supports dissociation, independently of arguments concerning the evolution of consciousness and attention.

Forms of attention without awareness have been found in memory research and also in studies of selective attention, as we noted in chapter 2. These types of attention are almost unanimously accepted in the psychological literature on attention. Any ideal theory of consciousness should make room for such findings on the very basic forms of attention, because although the information that is the subject of such attention does not satisfy the requirements of access consciousness, it is sufficiently detailed and is available to a subset of cognitive operations. Moreover, a comprehensive theory of conscious attention must be compatible with the scientific consensus that attention is an early adaptation that is shared by many creatures with very basic nervous systems (e.g., see Tooby and Cosmides 1995, Ward 2013). We develop this idea in chapter 5.

Research that specifically investigates the relationship between attention and consciousness shows that a dissociation must exist. As we discussed in chapter 2, there are a variety of situations where attention selects perceptual information to perform successful actions without the presence of conscious awareness. Many studies have shown that information processed by the basic forms of attention does not reach consciousness and yet can still affect behavior, decision making, and motor actions (e.g., Dehaene and Naccache 2001, Mulckhuyse and Theeuwes 2010, Norman, Heywood, and Kentridge 2013, van Boxtel, Tsuchiya, and Koch 2010). Blindsight is an extreme example of this type of cognitive ability to perform actions that rely on selective attention without being able to report seeing the object upon which the action is performed (e.g., see Brogaard 2012, Kentridge 2011, 2012). Ideomotor theory, which holds that action planning relies on having a perceptual representation of the result of the action (see Shin, Proctor, and Capaldi 2010), offers a plausible way to describe how some actions are performed; it is also perfectly compatible with severe dissociation, since much of motor planning occurs outside of conscious awareness.

Even those who favor a less severe (Type-A) dissociation suggest that there can be forms of attention that are deeply dissociated from conscious awareness. While some argue that attention is necessary to process information that becomes part of awareness (e.g., Dehaene et al. 2006), the conditions that are sufficient for conscious awareness still need to be addressed (e.g., Cohen et al. 2012). Koch and Tsuchiya (2007, 2012) implicitly support a Type-B or Type-C CAD: their belief that consciousness and attention are largely dissociated is based on various psychophysical and brain imaging studies on the neural bases of consciousness. As such studies suggest, there are neural activations associated with consciousness that are independent from those supporting attention; such findings, though not clearly conclusive, warrant further investigation. Whether brain imaging studies can truly clarify the underpinnings of consciousness continues to be debated, but as the technology and techniques advance, neurophysiological results can be expected to advance the clarification of the degree of CAD.

There are also good reasons to believe that consciousness is associated with a more integrative process than basic forms of attention. For example, experiments show that there can be access to semantic, numerical, object-based, and many other perceptual contents by means of attention, but that the level of neural processing required for *access consciousness* seems to require recurrent or feedback activations (e.g., Dehaene, Sergent, and Changeux 2003, Di Lollo, Enns, and Rensink 2000, LaBerge 2005, Lamme 2004, Raffone and Pantani 2010). This idea of consciousness serving an integrative function is related to the 'global workspace' view of consciousness (e.g., Dehaene and Naccache 2001). Such theories require us to examine the interaction of attention with other related cognitive systems, such as memory.

Crucially, Catherine Tallon-Baudry has tested the hypothesis that the contents of consciousness are identical to the contents of working memory, whose contents are considered by Jesse Prinz (2012) to be identical to the contents of attention. Consistent with our theoretical analysis, Tallon-Baudry (2012) says that the findings allegedly confirming the identity views are based on confused and biased approaches to data analysis. More important, she found neurological evidence suggesting a significant dissociation between the neural correlates of attentional access to information in working memory and of conscious experience. She proposes a model of consciousness and attention that entails a severe dissociation (at least a Type-B or Type-C CAD). She concludes,

> Experiments in which both attention and consciousness were probed at the neural level point toward a dissociation between the two concepts. It therefore appears from this review that consciousness and attention rely on distinct neural properties, although they can interact at the behavioral level. It is proposed that a "cumulative influence model," in which attention and consciousness correspond to distinct neural mechanisms feeding a single decisional process leading to behavior, fits best with available neural and behavioral data. In this view, consciousness should not be considered as a top-level executive function but should rather be defined by its experiential properties. (Tallon-Baudry 2012, 1)

Tallon-Baudry and her colleagues also found dissociations between voluntary and involuntary attention with respect to their interactions with conscious awareness (Hsu et al. 2011). A significant finding is that the data suggest a reverse interaction between attention and consciousness: unlike the typical view, which posits attention as the gateway to conscious awareness, their findings suggest the opposite and counterintuitive view, namely that awareness is what determines whether attention is voluntary or involuntary. In sum, dissociation seems to be confirmed by recent studies that look very carefully for ways of avoiding confounds.

Based on these findings, together with other findings on selective attention across species, it seems plausible to argue as follows for a Type-C dissociation: the neural correlates of consciousness and attention seem to differ more often than not. Even if the data are inconclusive, there are powerful theoretical reasons to favor dissociation: the distinction between integration, access, and phenomenality; and the need for the distinction between voluntary and involuntary attention. Tallon-Baudry's proposal to define awareness not as an executive function related to working memory, but only in terms of its experiential character, fits nicely with these theoretical constraints because executive function and working memory seem to be related primarily with what Block calls 'access consciousness.' We expand on this model below.

A different set of findings, mostly on mirror neurons, suggests a rather unexplored type of dissociation: conscious attention of the most integrated kind (phenomenal-access conscious attention, in which conscious awareness and attention overlap and interact) and *self-aware* conscious attention, perhaps the richest kind of conscious attention on the spectrum of integration.[22] Wolfgang Prinz (2012) argues that representation of the intentions of other agents, which requires social training and mediation, is one of the main functions executed by the mirror neuron system. But W. Prinz reverses the standard order of explanation, arguing that we first learn the behavior of others in order to learn how to interpret our own behavior and attribute its traits to our own intentions. In the following section, we explore the social dimension of conscious self-awareness and explain possible forms of dissociation.

3.4 Self-Awareness

Because of the centrality of epistemic agency with respect to access consciousness and its relationship to phenomenal consciousness, we devote a substantial portion of this section to further clarifying these distinctions, arguing for types of dissociation between self-awareness and epistemic agency. Our main conclusion is that epistemic agency does not necessitate self-awareness. This does not mean that epistemic responsibility is incompatible with non-self-aware epistemic agency, as epistemologists with internalist and coherentist convictions think. Rather, epistemic agency is in many cases dissociated from self-awareness, and remains the basis for many forms of epistemic responsibility.

A related and controversial issue is that discontinuities in what we called in chapter 1 the 'recognitional self' do not seem to entail discontinuities in what we called the 'phenomenal self.' Bayne (2010) makes a similar point by saying that the unity of consciousness (and subsumption) is compatible with discontinuities in the self. However, Bayne argues for this point based on a fictionalist characterization of the self, according to which selves are fully intentional entities, similar to the characters of a novel. Instead of making this claim, we will argue that Bayne's compatibility thesis is correct because of the general theory of conscious attention we defend, which favors a significant dissociation between phenomenal consciousness and self-awareness of the rational-recognitional type. Although our view also dissociates phenomenality from self-awareness, we believe our approach is incompatible with Bayne's because ours does not entail fictionalism.

In this section we defend five theses, in subsections 3.4.1 to 3.4.5. First, access normativity is epistemic normativity. Second, phenomenal normativity is necessary for self-awareness normativity. One can have phenomenal experiences without having self-awareness *recognition*, but the experiences would still be normative in a behavior-guiding way, as illustrated by the previous example regarding pain. The type of normativity required for self-awareness (i.e., metacognition combined with self-attribution) is based on phenomenal consciousness or the 'phenomenal self.' Third, phenomenal awareness is always *de se* awareness—that is, the type of awareness that involves self-attribution and self-location in the first-person indexical information sense, implying a minimal sense of ownership of one's own experiences and mental states. Therefore, recognitional self-awareness must be distinguished from *de se* phenomenal awareness. This is the distinction we made in chapter 1 between the recognitional and the phenomenal self.[23] Fourth, subsumption does not preclude that access and phenomenal consciousness can occur in tandem. In fact, subsumption is compatible with the systematic co-occurrence of phenomenal and access consciousness. This is compatible with Tallon-Baudry's cumulative influence model. Fifth, recognitional self-blindness is compatible with phenomenal self-experiences. This last thesis is compatible with Bayne's views on the unity of consciousness and the self. Additionally, we briefly address the issue of how the recognitional self is socially mediated, based on W. Prinz's (2012) theory.

3.4.1 Access Epistemic Normativity

In epistemology it has become increasingly important to distinguish between introspection-based epistemic processes and other forms of cognition that may produce justified belief. The debate has progressed toward more minimal forms of epistemic agency than those required by such strictly rationalist approaches as Descartes's. Reliabilism seems to be at the opposite end of the spectrum of views because it contends that aspects of the agent's internal mental life are not crucial for the evaluation of justified belief; a belief is justified as long as the *process* that produced the belief is more likely to result in true belief than false belief. So an important question is how to make sense of the epistemic responsibility of cognitive agents, given this spectrum of possibilities ranging from highly self-recognitional rationalization to a view that imposes no analytic requirements on the agent.

Some contemporary accounts of perceptual attention propose that only agents, epistemic agents in particular, can attend to features of the environment, and that the manner in which they attend is fundamentally dependent on cognitive activity of an explicitly epistemic kind (see Wu 2011). This idea was carefully developed by the so-called 'activity theorists,' who will be the main focus of our discussion in section 3.5. In this section we defend a minimalist view of epistemic agency, according to which one can be an epistemically responsible agent without having phenomenal consciousness: access consciousness suffices for epistemic responsibility. Indeed, we shall argue that access consciousness is deeply associated with epistemic normativity. We first show that access conscious normativity is epistemic normativity, and then that the most plausible interpretation of the normativity inherent in phenomenal consciousness is that it is deeply related to social empathy. Thus, in the light of empirical findings, dissociation is the most plausible interpretation of these two kinds of normativity.

For example, a super-duper-blindsighter, like Maria, may be a flawless epistemic agent, but she will not be able to *empathize* with our experiences. One can have phenomenal experiences without self-awareness of the recognitional kind, but one cannot have self-awareness without phenomenality.

We have been arguing that there is considerable empirical evidence for a dissociation between attention and consciousness. If this is true, then such a dissociation will also manifest in a *normative* dissociation. To defend this view we must identify what *type* of agency is required for epistemic achievements and knowledge attributions, and that identification requires an epistemic analysis. Because agency is fundamental to our account, we focus on reliabilist versions of virtue epistemology to address this issue.[24]

Some reliabilist theories of epistemic virtues require a reflective component for knowledge.[25] For instance, Ernest Sosa (2009) distinguishes between two types of epistemic agency, a nonreflective and a reflective type. Those yield two types of knowledge attributions: animal and reflective knowledge. According to Sosa, it is only when agents achieve reflective knowledge that they display the highest, most reliable and apt form of knowledge, full knowledge. Reflective knowledge seems to involve both metacognition and metarepresentation. It allows for a refined type of epistemic achievement in which object-based, first-level 'animal knowledge' is assessed in terms of the way in which such knowledge was produced, taking into consideration the overall epistemic situation of the agent. Thus, reflective knowledge concerns the overall adequacy of the belief given the epistemic situation of the agent, an assessment that animal knowledge cannot provide.[26]

In contrast, John Greco (2004, 2010) argues that a reflective requirement on knowledge imposes unrealistic psychological demands on epistemic agents. One can understand this claim as follows. Sosa's account of full knowledge introduces distinctions where there are none. For instance, when a child accurately perceives a red object and forms the belief that there is a red object in front of her, she seems to *know* that there is a red object in front of her, and we *should* attribute such knowledge to *her*, because this knowledge is an achievement of her reliable perceptual skills. Sosa grants that this is a case of knowledge—animal knowledge—but the question is, why would a reflective component be required for her to *fully* know that there is a red object in front of her when she seems to know everything there is to know about the case?

It seems that reliably detecting features of the environment may be only one of many epistemic achievements that are required for full knowledge. Sosa illustrates this with Diana, the huntress, who has to be not only successful in hitting targets, and reliably so, but also *selective* with respect to the type of target she *should* hit. So not just any metacognitive process will do. For instance, just monitoring perceptual processes will not suffice. Selection based on *criteria* seems to be an important epistemic achievement, which is fundamental to assess evidence and withhold judgment. This is what Sosa calls 'reflective justification.' Based on psychological terminology, the reliable object-based knowledge that Diana uses to succeed in hitting targets constitute her 'motor control' skills, and the reflective knowledge that she uses to select appropriate targets constitute her 'action selection' skills.

What is epistemic 'reflection'? Clearly it involves control, selective criteria, and a higher degree of voluntary involvement on the part of the epistemic agent; certainly more involvement than successful behavior based on motor-control ability requires. Maria, the super-duper-blindsighter, seems to comply with all these requirements. Does she also have full conscious access to normative criteria for selection? Here one must avoid confounds. In an obvious sense, Maria is just like Diana and can be as selective. But requiring full *phenomenally* conscious control of the agent is not the same as requiring epistemic access and *selectivity*. Greco seems to be right in his criticism that reflective justification, understood this way, seems an excessive requisite for epistemic reflection.

Hilary Kornblith (2010) argues that reflection is a problematic requirement for knowledge because instead of increasing reliability and success, as Sosa seems to think, it actually *reduces* or *interferes* with them. This may be one aspect of a larger problem, which is that the reliability of motor control skills is largely *dissociated* from action selection and thereby from phenomenally conscious reflection. How minimal can the conscious cognitive requirements for epistemic responsibility be? We shall argue that one can dispense with phenomenality and preserve most forms of epistemic responsibility. In section 3.5, we discuss a form of agency in which there is phenomenality with no experience or sense of self: the experience of 'flow.'

3.4.2 Epistemic Agency, Success, and Access

Many cases of knowledge involve the success of epistemic ability in such a way that the connection with action is obvious. When you are driving to an important meeting and your copilot insists that she knows how to get there, but keeps getting it wrong, it seems you should conclude that she doesn't know how to get there. In such a case, it is best to rely on a computerized navigational system like a GPS. Standard *process* reliabilism about knowledge has no trouble accounting for the type of knowledge one achieves by following the information from the GPS. But cognitive integration alone, even if it leads systematically to epistemic success, is not sufficient for epistemic agency. Access consciousness, by contrast, is sufficient: one needs a reliable perceiver and interpreter of GPS information with basic epistemic capacities to draw proper conclusions from its information.

How much phenomenal introspection is required to succeed? It seems that paradigmatic cases of perceptual belief involve very little introspection, in the sense that perceptual abilities give agents immediate knowledge of the environment without reflective, phenomenally conscious effort. The same is true about the navigational abilities that produce spatiotemporal knowledge. The perceptual-like reliable abilities of an agent seem to satisfy the requirements for knowledge and epistemic normativity, even if no phenomenology is involved. These are the abilities that constitute *motor control* skills.

Diana the huntress uses motor control skills to hit targets with her bow and arrow. She forms immediately justified beliefs, or at least epistemic entitlements, about the position of the targets (from basic perception), and through skillful motor control she deploys her knowledge of how to best balance the tension of the bow, taking into account atmospheric conditions and distance (from specialized knowledge based on memory). She can do this almost mechanically. All this knowledge about archery is attributable to her because it is nonaccidentally based on her skills, not dependent on luck or other contingencies of the environment. If Diana were a good archer because someone else were always helping her, or because she had so far been incredibly lucky, having always favorable winds, then she would lack such knowledge, and we should not attribute such knowledge to her.

Psychologists have confirmed that perceptual and motor control skills are very reliable. Perceptual knowledge is based on stable dispositions to respond to stimuli across similar environments. The same is true about procedural knowledge and forms of implicit memory, which require a very minimal type of agency; these types of knowledge can be achieved unconsciously. What is important to highlight is that knowledge for motor control is reliably produced by abilities that are largely attributable to agents but that don't require the agents to be phenomenally conscious of how they perform these epistemic processes.

Diana is not only an extremely reliable archer; she is also a superb hunter. Her performance as a hunter cannot be fully captured by her first-level reliability as a good archer, because formidable hunting performance must manifest her selective judgment regarding appropriate targets. Hitting just any target would be detrimental, waste energy, and reduce first-level reliability by introducing an eagerness to hunt that may be incompatible with precise archery. She must hit specific prey and refrain from hitting anything else, thereby displaying the adequacy of her decisions. We would not attribute this reflective knowledge to Diana if she decided to hit prey by flipping a coin or by basing her decisions on the instructions of someone else. Thus, her knowledge for action selection is a second-level type of luck elimination for epistemic goals, attributable to her because of her selective skills. As we explained above, mere cognitive integration does not suffice for epistemic agency, and the same holds for reliability: mere reliability based on cognitive integration does not suffice for epistemic agency.

It seems clear that the reflective skills that Diana manifests, and her overall epistemic situation with respect to prey, depend critically on her voluntary intervention. One could think of these skills as highly procedural; she may detect targets and reflect on the adequacy of shooting prey quite quickly and by means of implicit knowledge and expertise. The evidence shows that such skills cannot be integrated into stable, systematic, or coherent epistemic virtues, if by 'voluntary intervention' we mean *phenomenally conscious* introspective deliberation.

How far up the cognitive integration and epistemic access hierarchy does one need to go to find the type of agency required for reflection? Most examples of reflection involve what seem to be descriptions of conscious deliberation. Diana can, however, hit targets very rapidly by relying on epistemic processes of which she is unaware, which constitute her motor control knowledge. But this is not a plausible characterization of her reflective, *action selection* processes. She is clearly involved in a more robust and conscious way in the selection of such actions as hitting all and only appropriate targets than in the act of just hitting targets based on her perceptual knowledge. So how to describe this second-level enrichment of her archery skills without characterizing it in terms of phenomenal introspective judgment?

In the next section we propose a solution to this problem, based on current accounts of ideomotor theory, and explain how Nicholai Bernstein and the activity theorists addressed the issue. The key aspect of our proposal is that all enrichments concerning expertise can occur without phenomenally conscious monitoring by a central executive function. But even if one favors a central-executive reflective account, we argue that it is *access conscious* execution that is required, not phenomenal experience.

It is useful to think of this problem in terms of knowledge attribution. Diana clearly knows how to hit targets because of her first-level knowledge, and without the second-level knowledge her hunting would be deficient. Knowledge for hunting should not be attributed to her in the absence of action selection skills. We claim that whatever is true about Diana concerning knowledge attribution should be true about Maria, the super-duper-blindsighter, because of her high-quality epistemic access to contents and her capacity to be highly selective in both motor control and action selection. This suggests that very high quality epistemic access is constitutive of epistemic agency without necessitating phenomenal consciousness. Access consciousness captures exactly this idea, with the addition that contents are available for thought, report, action, and memory.

Kornblith (2010) suggests that a diminished reliability results when there is an interaction between phenomenal introspection and motor skills. He refers to introspective forms of reflection about one's own epistemic status, which have been proven to be highly unreliable.[27] The proponents of reflection may respond by saying that these processes are too high up the scale of reflection, and that more minimal forms of reflection could do the job. In the next section we show that this response is problematic because any form of action selection construed in terms of introspective phenomenal consciousness, no matter how minimal, can be epistemically dissociated from motor control and perceptual-like knowledge, which impedes the kind of cognitive integration required for epistemic responsibility. (This dissociation is related to those between consciousness and attention.)

3.4.3 Success and Cognitive Dissociation

The distinction between motor control and action selection is crucial for understanding why phenomenal introspection is epistemically problematic. Action selection for basic tasks, like choosing one option over an alternative when one is *inclined* to do both, seems to be the most adequate candidate for introspective reflection. Findings on action selection suggest that the richer the sense of agency, the less epistemically relevant the characterization of introspective phenomenal processes; a rich sense of agency does not correspond to epistemic benefit. The problem is not just one of diminished reliability, which could be remedied by finding other forms of cognitive reflection characterized by fewer introspective constraints and greater connection with perceptual processing. The fundamental issue is that phenomenally introspective action selection seems to be epistemically dissociated from reliable belief for motor control.

To illustrate the distinction between motor control and action selection, it is useful to start with a nonperceptual case, or at least one that is not strictly perceptual, so as to demonstrate the generality and importance of this distinction. Knowing how to speak a specific language, for example, requires knowledge of its syntax. It also requires knowledge of how to articulate and generate the specific sounds associated with its letters and words. Syntax processing and other motor-control aspects of linguistic representation are beyond our phenomenally conscious introspective reach. Syntax is the systematic manipulation of information in terms of the strictly formal characteristics of linguistic stimuli. A string of symbols or sounds are processed according to formal distinctions such as subject, predicate, noun, adverb, and so forth. Meaning is embedded, modified, and composed according to these rules. This is not only *phenomenally* unconscious cognitive processing, it is also strictly sensory-motor knowledge, and no form of reflection needs to occur for speakers to have this knowledge. In speech production, one is phenomenally conscious of the meaning of words, but not of the syntactically driven articulatory code that is involved in articulation and sound intonation.[28]

Knowledge of syntax is a crucial ability that underlies knowledge attributions concerning language. Speakers know a language *because* of this ability. An aspect of syntax processing that is crucial to understanding the importance of motor control is its reliability. Knowledge of syntax concerns the type of reliable process that leads to immediate perceptual beliefs. Just as seeing a tree leads to the kind of perceptual knowledge that requires minimal reflective capacities—knowledge about the identity of objects—knowledge of syntax is dependent upon reliable processes that require minimal agency and no kind of reflection or monitoring. According to Sosa, knowledge of syntax should qualify, oddly, as animal knowledge even though it seems to be a uniquely human trait.

One may object that this challenge to reflection is theoretically biased toward an impoverished and unstructured characterization of epistemic achievements. Unconscious, subpersonal, and mechanical accounts of knowledge that are compatible with this proposal are minimalist views that most epistemologists would find unpalatable. This objection loses its force once one considers carefully the distinction between different types of cognitive integration. It is true that modular-like processes are inadequate to account for the type of unity characteristic of phenomenally conscious integration, or of higher-order processes that involve metarepresentations of a unified self, although even these processes do not constitute a unique type of integration (see Proust 2007). But these processes are sensitive to information that is *epistemically irrelevant*, such as how to best interpret information to achieve phenomenal integration, or how to best satisfy pragmatic goals, rather than achieving such epistemic goals as producing true belief.

As Greco (2010, 166) argues, cognitive integration for epistemic abilities must be sensitive *only* to those parts of the cognitive system that produce reliable information, and *insensitive* to those that produce unreliable information or play a different, not necessarily epistemic, role. His example is that perceptual beliefs are insensitive to highly theoretical beliefs such as philosophical beliefs about the inexistence of the external world, or beliefs about particle physics. Likewise, they must be insensitive to the phenomenology of reflective introspection.

This kind of epistemic sensitivity is captured by the psychological notion of *cognitive impenetrability* (see Pylyshyn 1980). In many standard cases of color, shape, size, and depth perception, as well as cases concerning other properties of perceived objects including Gestalt effects, background knowledge and emotion seem to affect very little how one perceives these constancies. Thus, integration just for the sake of reflection need not be epistemically fruitful, and it is not necessary for epistemic agency.

Even in the case of perceptual belief, not all of the cognitive integration for perception is epistemically sensitive. Stöttinger and Perner (2006), for example, showed that motor control is not influenced by an illusion, unlike phenomenally conscious perception. One can phenomenally see the difference in the lengths of two lines in the Müller-Lyer illusion, even though one knows (and therefore truly believes) that they are the same length. Our phenomenally conscious visual perception is, in this particular case, impervious to reliable epistemic influence. Surprisingly, motor control is epistemically sensitive to such reliable information, even in cases of perceptual illusion. In other words, motor control has the correct epistemic access that is available for non–phenomenally conscious action and cognition. In the Müller-Lyer illusion, although the observer's conscious self-report is inaccurate and reflects the illusion's cognitive influence, her motor control, specifically her unconscious manual behavior for grasping, is accurate and not influenced by the illusion.

In their experiment, Stöttinger and Perner presented subjects with vertical lines grouped in two sets, one with open brackets and the other with closed brackets, as in the standard Müller-Lyer illusion. When asked "which gang of lines would you fight?" subjects chose the "smaller" lines although their motor control when reaching for those lines in the absence of this question did not distinguish between the sets of lines because it was not influenced by the illusion. This finding demonstrates a dissociation between action selection and motor control.[29] Conscious inclinations about fighting are clearly not fundamentally associated with *epistemic reliability* and, in this particular case, the conscious decision to fight the smaller lines is based on false information. This information for action selection may lead to good practical decisions, but not to true belief. Accurate motor control concerning length, on the other hand, is a precondition for successful navigation. So it makes sense that the epistemically relevant information that allows agents to succeed based on their knowledge of the environment would ignore or be insensitive to the epistemically irrelevant phenomenally conscious information concerning whom to fight. Cognitive integration for motor control processes that lead to reliably successful actions is insensitive to epistemically irrelevant inclinations and to highly sophisticated theoretical or philosophical beliefs, in spite of the fact that those inclinations may underlie practical and *socially relevant* interests. However, as the example just mentioned shows, cognitive integration for phenomenally conscious processes and action selection *is*, in some cases, sensitive to epistemically irrelevant information.

For the purpose of our discussion we can now reach an important conclusion. Since epistemically sensitive cognitive integration is a precondition for access consciousness, access consciousness seems to suffice for explaining any case of knowledge attribution. Epistemic normativity is access consciousness normativity (not phenomenally conscious normativity), and epistemic reliability is best understood in terms of access consciousness. Attention is epistemically sensitive to information and integrated in terms of access consciousness. This conclusion also supports a dissociation between phenomenal consciousness and attention, with the innovative claim that the dissociation manifests normatively. We now proceed to other issues related to self-awareness and reflection that support this conclusion.

3.4.4 Subsumption, the Self, and Self-Knowledge

The intuition that one remains the same self throughout one's life is an extremely powerful one. About this intuition, James (1890/1905, 317) wrote, "Each of us, when he awakens says, Here's the same old self again, just as he says, Here's the same old bed, the same old room, the same old world."

In light of the arguments we have introduced so far, however, this powerful intuition can be interpreted in at least two ways. One interpretation is that the self one experiences as invariant throughout one's conscious life is the phenomenal self. This experience occurs without self-knowledge, metarepresentations, or cognitions. It is a fundamental feature of phenomenality that one experiences phenomenal contents as essentially integrated to one's own first-person perspective, or phenomenal self.

An alternative interpretation is that the self James is talking about is recognized—by some epistemic process, inference, or categorization—as the same self because of a judgment concerning identity through time. We argue that while both interpretations are plausible, only the first gives rise to the powerful intuition James discusses.

In support of this claim we defend the following thesis: phenomenal normativity is necessary for self-awareness normativity, including capacities for empathy. Although one can have phenomenal experiences without having self-awareness *recognition*, these experiences are normative in a behavior-guiding way, as the above example about pain illustrates. Phenomenal awareness seems to always be *de se* awareness, in a nonrelational and noninferential way. This seems to be a more primitive way of understanding *de se* representations, because phenomenal awareness is constitutive of perspective and *de se* independent of beliefs and their relations to facts. So it seems that recognitional self-awareness must be distinguished from *de se* phenomenal awareness. We shall also argue that the unity of phenomenal consciousness does not preclude that access and phenomenal consciousness may occur in tandem; and that self-blindness of the recognitional type is compatible with phenomenal self-experiences. This is a plausible way to interpret Tallon-Baudry's cumulative-influence model between attention and consciousness, as well as Bayne's views on the unity of consciousness and the self. We first defend the normativity thesis, then address the dissociation thesis concerning recognition and self-knowledge.

The knowledge Mary gains after her release into the world, based on Frank Jackson's (1982) famous argument, helps illustrate the type of not strictly epistemic normativity that, we claim, characterizes phenomenal content.[30] When Mary, a neuroscientist who has scientific and physical knowledge of color, is released from her colorless environment and experiences red for the first time, it seems that she gains knowledge concerning what it is like to see color that she did not have before. Since this knowledge cannot be (by assumption) physical, third-person based, or scientifically based knowledge, it seems that the knowledge she gains is of a nonphysical nature. This new information she acquires, we suggest, is not of a strictly *epistemic* kind, although it can be correlated with strictly epistemic information. But what can this mean?

There are three options for interpreting the claim. One is that what Mary acquires is epiphenomenal information: nonphysical (i.e., not causally relevant) information about properties of her experience. No explicitly normative thesis can be derived from this largely negative claim, particularly with respect to epistemology. More specifically, it is unclear how justification and truth, which drive what is epistemically normative, could characterize strictly phenomenal information. Moreover, the causal nature of knowledge—the causal-evidential connection between nonaccidentally justified beliefs and the truth—also makes this option highly problematic.

A different interpretation is that the normativity of the phenomenal information Mary gains cannot be epistemic because she already knows everything there is to know about color. But then it seems that Mary not only knows everything, she also understands everything about color and no added normativity of any sort occurs. This is what Chalmers (2003a) would call the standard type A materialist response to Mary's predicament, and it leaves little room for explaining what kind of cognitive enrichment Mary undergoes.

Our proposed account of Mary's situation, following David Lewis's (1988) ability hypothesis, belongs to the family of responses that grant that

Mary learned something, without granting that what she learned involves new facts. The point we want to make is neither metaphysical nor semantic. Rather, we want to illustrate how the dissociation between phenomenal consciousness and attention may entail a normative dissociation.

The general idea is that what Mary learns is the same facts she knows about color, but under a new 'mode of presentation' (to use Frege's term), which allows for *empathy-normativity*.[31] If you taste vegemite for the first time, or have any other experience for the first time, even if you know all the physical facts about it, you seem to gain an insight that goes beyond such knowledge. You can think about how other people's experiences *may* resemble yours; you can wonder how anyone could possibly enjoy the taste of vegemite. The experience of red is an ability new to Mary, but it is clearly not a strictly epistemic ability, because she does not acquire new knowledge. Her cognitive enrichment is best described as a new ability to empathize with others based on knowledge she already has. What Mary learns seems to be a form of understanding, a form of empathy.

Given our previous conclusion that access normativity is epistemic normativity, one has to conclude that although what Mary learns cannot be epistemically normative, it has to be a different normativity. This would be considered an empathy-based kind of normativity, of the sort which presumably grounds aesthetic and moral values. One way to interpret this proposal epistemically is that *understanding* based on empathy is a different achievement from *knowledge*, because it provides a different mode of presenting already known facts. The kind of empathy based on phenomenal consciousness can be cognitively integrated with epistemic information, but this requires the involvement of the functions of access consciousness. The key is that access consciousness provides reliable information and is epistemically sensitive to accurate information, including the *origin* of such information.

This function of access consciousness can be illustrated with research on reality monitoring. According to Marcia Johnson (1991), because of the number and variety of sources of information the brain must integrate and analyze, two processes are critical in identifying the reliability and source of information. She calls these processes 'reality testing,' which applies to ongoing information, and 'reality monitoring,' which applies to memory. We will call both of them 'access monitoring,' because the goal is for the brain to know the causal origins of information. In most illusions where there are accuracy conditions that are being misrepresented, subjects know that the information is coming from a perceptual source, somewhere outside their skull, as it were. But in cases of hallucination this monitoring fails, and subjects cannot recognize that the information perceived is coming from within, so they attribute it to external sources.

Perceptual afterimages are interesting precisely for this reason. Subjects attribute properties that are generally attributable only to external visual stimuli to stimuli that they *recognize* as internally produced. One can say, "I am experiencing an afterimage that looks red and square, but I *know* it is not there and that it will disappear soon." Although one is not mistaken, it is puzzling that one still assigns properties (color, shape) that can only be attributed to external stimuli. The language of phenomenal consciousness is used to refer to stimuli that are correctly identified as epistemically inadequate. At the same time, our perceptual attention for navigation is not fooled by afterimages. If one were asked to grab the red square one sees, one would laugh or find such a request incoherent.

Our claim is that access consciousness and attentional capacities for perceptual tasks include this kind of monitoring. Access monitoring is, therefore, epistemically normative. Phenomenal understanding for empathic abilities, however, is not epistemically normative, not least because it can be dissociated from access monitoring. In dreams, hallucinations, and confabulations, we do not have such monitoring, and attention cannot work properly; but we do have phenomenal awareness and are emotionally and aesthetically responsive to the contents of phenomenal consciousness. Crucially, we are not *epistemically* responsible in such conditions. In such situations phenomenal consciousness runs free. In dreams we experience the same pressure to please others as in waking life, the same need to avoid fighting intimidating gangs, and the same pleasure when we see a beautiful vista or the smile of a fond friend. But there are no epistemic constraints on these cognitions. In dreams we understand the pain of others and their moral requests. We also have the *conviction* that what we see is real, that what we perceive is "out there." But the reality monitoring system, along with the motor control system, is turned off. This reinforces our theoretical claims concerning the dissociation between phenomenal conscious attention and access conscious attention, and supports our theses about access and phenomenal normativity.

3.4.5 Recognitional Self-Awareness

Although one can have phenomenal experiences without having self-awareness *recognition*, such experiences are normative in that they can affect our behavior. A creature can have an experience of pain, and react to it, without being able to *recognize* the experience as something that is happening *to it* as a subject of a unified and temporally extended experience. The experience will be able to guide the creature to avoid situations that produce pain, and such a response does not require any knowledge of what pain is. The experience, as we shall argue below and in the next chapter, will also allow this creature to understand pain as something that other creatures may have.

While the same capacity for access monitoring with respect to sources of information seems to extend to attributions of contents to oneself (e.g., judgments concerning our own cognitive capacities), such attributions are distinct from the phenomenal qualities of experiences. Bayne (2010, 290) says that failure in a subject's ability to track her own thoughts and experiences does not entail that these experiences are not *de se*, that is, fundamentally experienced by the subject as something that is happening *to* the subject. Paraphrasing Bertrand Russell, one never experiences a sunset as an "impersonal" experience that may be happening to someone else. What is troubling about schizophrenia, for example, is that one attributes the sources of one's own thoughts to someone else; but even that should not entail that those experiences lack the subjectivity or minimal sense of ownership that defines them. Thus, the phenomenal self can be dissociated from the recognitional self because the minimal sense of ownership that characterizes having an experience does not require epistemic monitoring of sources of information, and this entails another kind of dissociation with respect to phenomenal conscious awareness: a dissociation of phenomenal self-awareness and recognitional-phenomenal self-awareness. The type of normativity required for self-awareness of one's own experiences, whether metacognition, self-attribution, or putative mind reading, is essentially *phenomenal*. What access monitoring provides is epistemic verification of the sources of phenomenal experiences. Since phenomenal awareness is always *de se* awareness, then recognitional self-awareness must be distinguished from the phenomenal self. Recognitional self-blindness is compatible with phenomenal self-experiences because lacking the capacity to recognize the source of an experience one is having does not entail that one is not experiencing it subjectively.

Moreover, subsumption does not entail that access and phenomenal consciousness cannot occur in tandem. Subsumption integrates different experiences with their own degree of empathy-oriented normativity, and when these experiences are properly monitored in waking life they provide evidence of contents. Such contents also become epistemically normative by the interaction between motor control, causally driven attention, and phenomenality.

To conclude, then, understanding others seems to be deeply related to understanding oneself *phenomenally*, and this seems to mean a capacity for having phenomenal experiences is needed in order to understand the experiences of conspecifics. Empathy, perspective taking, and imagination go beyond the capacity to know and access contents; they involve abilities that are not merely epistemic, but seem to give rise to moral and aesthetic value. When they occur in tandem, these capacities generate contents that can be monitored for sources of information, self-attribution, and other complex tasks. This comports well with the thesis that the phenomenal consciousness associated with these capacities could not have evolved as an early adaptation, an issue we expand in chapter 5.

3.5 Social Influence and Activity Theory

We conclude our discussion of theories of consciousness with some remarks about the implications of the dissociation between phenomenal and access conscious attention for socially mediated cognition. Phenomenal consciousness, when it is not monitored, may go epistemically astray, but would still provide a basis for moral and aesthetic evaluations. One may be delusional, or be dreaming, and still be appalled at someone's pain or feel the joy of holding a loved one. Phenomenal character, as mentioned, opens our mind not merely to the world as a series of facts, but also to how other people *might* experience the world. In this sense, phenomenal character is socially fundamental.

W. Prinz (2012), partly relying on research concerning mirror neurons, suggests that the capacity we have to engage in imagination and planning from our own perspective critically depends on how we understand the experiences of others. If W. Prinz is right, phenomenal character may also be fundamentally social. The notion of 'mirroring,' however, also requires disambiguation. An epistemic agent who cannot experience pain but knows a lot about pain can mirror the behavior of someone who is in pain. She could mirror patterns of behavior and perhaps even know how to predict certain outcomes, but she could not fully empathize with the situation of those in pain. This holds for any experience, including those of patients with Urbach-Wiethe disease, who cannot experience fear. Prior to her release, Jackson's Mary cannot understand what people feel when they see red surfaces. Even if it could be hallucinated in Mary's case, only the actual experience of color opens the possibility to empathize. If this is correct, phenomenal consciousness is *normatively* distinct from access consciousness. We can call it *empathic* normativity. If it provides epistemic normativity, it does so only insofar as it occurs in tandem with access consciousness.

Self-awareness and self-knowledge are crucial for explaining the experience of flow, or effortless attention. As social creatures, the enrichments that refine our immediate reaction to external stimuli are determined by training and social convention, as is evident in cases of expertise. Before an action set (a complex set of goal-oriented actions that involve motivational and perceptual components) is automatically triggered by a specific stimulus (e.g., a signal by a teammate in a basketball game), previous monitoring and a focused voluntary attention, with significant effort, are required to produce a skilled response.

Social cues include not only those that would trigger action sets for collective action, like playing in an orchestra, coordinating a pass, or participating in a hockey play, but also less outcome-specific and more ritualized actions, like boarding an airplane in an orderly fashion, greeting a friend versus greeting your boss, or participating in a religious ceremony. The degree of effortful monitoring that these actions require correlates with the trade-off between unlearned and automatized behavior for action sets. Effortless attention, then, seems necessarily to be a form of *conscious attention*—an overlap between phenomenal consciousness and access conscious attention—because the phenomenal experience of subjective effort disappears even when the complexity of the cognitive task is significant. This entails a substantial amount of high-quality epistemic processing, intense engagement in the task with the concomitant experience of pleasure or flow, and little or no self-awareness.

When combined, phenomenal and access conscious attention provide a powerful way to interpret one's own behavior in terms of the experiences and expectations of others, and to identify cues that trigger action sets that rely on knowledge of the environment. A model that proposes such 'in tandem' relations between consciousness and attention is Tallon-Baudry's (2012) cumulative influence model, based on the distinct neural correlates of phenomenal conscious awareness and attention. We propose that this neural dissociation is best interpreted as the 'in tandem' functioning of phenomenal and access consciousness, that is, the interaction between consciousness and attention.

A potential consequence of W. Prinz's mirroring proposal, understood in terms of the dissociations we have discussed, is that it may provide an empirically adequate way to find a balance between Lewis's (1969) view on the nature of social convention—which emphasizes rationality, belief, and mutual expectation—and Ruth Millikan's (2005), which dispenses with rational constraints and emphasizes reproduction. The main idea behind our interpretation of W. Prinz's proposal is that in addition to the social practices that go into mimicking and mind reading conspecifics, these activities have cognitive components that ground normative judgments. One type of grounding is epistemic and the other is based on empathy. While interpersonal mirroring seems fundamental to enrich the action sets we use in our daily behavior, it must also guide action, and thus is normative in some sense.

Crucially, these mappings for action sets concerning phenomenally and access conscious attention can occur without recognitional self-awareness. They may even be reflex-like, triggering and engaging various forms of low-level attention, such as visual routines. Commenting on Pavlov's pioneering work on conditioned reflexive training, Bernstein (1950/1996) says that *any* enrichment of our personal experiences, particularly with respect to dexterity, like piano playing, deep-sea diving, attending a meeting, or wrestling, must occur in accordance with a complex and layered structure of reflexes, all driven by specific stimuli and motor tasks or action sets. In fact, according to Bernstein, these enrichments of conscious states have to be reflex-like, because successful performance requires fine adjustments and automatic responses triggered by the conditioned reflex's stimuli.

External cues that trigger attention-based action sets can facilitate and expedite action selection, reducing significantly the need for complex comparisons among an incredibly vast number of possible (and often mutually incompatible) responses. Had these refinements and cognitive enrichments necessitated step-by-step phenomenally conscious introspection, our cognitive capacities would be very inflexible and slow. Activity theorists like Bernstein, Dobrynin, and Leontiev developed systematic theories of the organization of attention and have proposed several types of attention, including one called *postvoluntary attention,* that is specifically related to flow experiences—a notion that itself is based on the work of Bernstein (see Dormashev 2010). Aleksei Leontiev's theory of activity (1978) makes goal selection in relation to specific cues the kernel of psychological analysis. Postvoluntary attention seems to be the most complex combination between phenomenal and access attention *without* highly introspective or self-recognitional capacities. These activity theories are compatible with, and actually reinforce, the dissociative account we have defended in this chapter.

Yuri Dormashev (2010) interprets Leontiev's theory of action and attention as an "activity Gestalt" account, which seeks to clarify the relation between conscious content and automatic responses to action sets. The activity theorists seem to have anticipated the important theoretical insight that goal-driven selection is crucial to understanding the mechanisms underlying attention, which mediate between unconscious automatic behavior and conscious (phenomenal as well as access) goal selection. This insight has played an important role in contemporary analyses of systematic associations between external cues for action selection and task switching (see Hommel 2010, 134–6).

Because of the analogy between conscious action selection and the type of conditioned or trained automatisms explored by the activity theorists, a topic that has recently generated interest is the notion of effortless control. The intuitive account of the cognitive effort associated with voluntary attention and monitoring is that the higher the demands of a task, the greater the effort associated with executing it, particularly with respect to sustained efficacy as the action unfolds in time (see Kahneman 1973). But in flow experiences, one finds the opposite correlation: when fully engaged as a trained athlete in a competition, a concert pianist, a master chess player, or a professional writer (like our friend Lloyd), the task demands increase, but the effort experienced *decreases* or remains constant (Csikszentmihalyi and Csikszentmihalyi 1988). In these circumstances one has little or no awareness of the recognitional self, while the phenomenal sense of self seems to be enhanced: the phenomenology of experience seems to be richer.

Although the notion of effortless control for highly demanding tasks is counterintuitive, the experience of flow is familiar. One forgets how difficult it is to perform a highly skilled action if one is very well trained to do it. In fact, the experience of executing a task that one is trained to perform with a high level of precision is one of enjoyment and relaxation, as research on these experiences documents (Csikszentmihalyi and Csikszentmihalyi 1988). More important, the demands and cognitive costs of the task are not *introspectively transparent* in cases of effortless control, which is characteristic of flow experiences. How is this possible, given that the attentional burdens are very demanding and context sensitive, while cognitive resources are limited? One possibility, which we have argued for in this chapter, is that these tasks do not require 'central' or constant executive monitoring, but rather, a complex network of reflex-like relations between specific stimuli and action sets. No single reflective and inferential self seems to be in charge of these highly sophisticated tasks.

3.6 Summary

Regardless of the normative dimensions of the dissociation between phenomenal consciousness and attention, there are powerful analytic reasons, both theoretical and empirical, to believe the dissociation exists. The distinction between phenomenal and access consciousness, higher-order theories (including those that reject the distinction between access and phenomenal consciousness), and considerations concerning the unity of consciousness all entail some form of dissociation between consciousness and attention. Moreover, epistemic considerations and issues about agency also strongly favor dissociation. Finally, an analysis of the role of self-awareness capacities involving recognition and socially mediated behavior also points to dissociation; research on effortless and effortful attention lends further support.

Collectively, the arguments presented thus far make a strong case for dissociation and pose a significant challenge for identity views. Even if some of these arguments were to prove inadequate, the case for dissociation—for at least a Type-B CAD—would remain substantial.

In the following chapter we continue our defense of CAD, expanding on the topics that have been covered so far, especially object-based conscious attention, Gestalt contrasts, emotions, memory, dreams, self-awareness, and perceived agency.

4 Conscious Attention

In previous chapters we have argued that there are different forms of attention with distinctive functions, and that several views on the nature of consciousness entail a form of dissociation between attention, defined functionally as epistemic access to information, and conscious experience. Taken collectively, these arguments, which are based on our current theoretical and empirical understanding of consciousness and attention, strongly suggest that there is a dissociation between consciousness and attention. In this chapter we examine the theoretical possibility that systematic forms of overlap exist between conscious experience and attention—the overlap is what we have termed 'conscious attention.' This possibility is compatible only with views that dissociate consciousness and attention without denying that they can overlap in regular ways. The views that preclude such an overlap are identity theories and full dissociation theories. For identity theories, by assumption all forms of consciousness are automatically forms of attention. At the opposite end of the spectrum, full dissociation theories allow for no overlap between consciousness and attention; although they might seem to occur in tandem, such theories must claim that there are no systematic overlaps between them.

For those theories that distinguish access consciousness from phenomenal consciousness, which we argued imply some kind of dissociation, the question is how to explain overlaps between consciousness and attention. The same problem arises for views that distinguish conjunctive epistemic access to mental contents, which is associated with attention, from the unity of phenomenal consciousness. A problem with this latter view, as we mentioned in the previous chapter, is that it is plausible to interpret it as entailing full dissociation. Views that reject the notion of phenomenality might still make a distinction between global access attention, which global workspace theories would consider as cases of conscious attention, and subpersonal forms of attention. In fact, these views are in principle compatible with further distinctions from the point of view of the nature of cognitive integration, such as the distinctions between unconscious attention, conscious attention, and preconscious highly integrated attention.

Given that attention is functionally characterized, and that many accounts of phenomenal consciousness deny that it can be described functionally, proposing a theoretical categorization of overlaps between consciousness and attention is not a trivial matter. We pursue this issue in chapter 5, based on evolutionary considerations. In this chapter our goal is to illustrate the type of cognitive achievement that would count as paradigmatically characteristic of conscious attention, assuming that such a category actually designates a natural kind, rather than a cluster of distinct processes. Thus, we shall neither adjudicate whether conscious attention exists, nor conclude whether or not it is a single type of cognitive process. We shall, however, argue that consciousness and attention likely overlap, at least in the cases we will discuss. One reason why we nevertheless cannot decisively conclude that 'conscious attention' refers to a natural kind is because of the likelihood that consciousness and attention evolved separately, which we examine further in chapter 5. When considering evolution, we are confronted with perhaps the most decisive argument against identity views. Those considerations favor some degree of dissociation of consciousness and attention, though without entailing full dissociation.

Another complication regarding the overlap or interface between the functions of attention and phenomenal experience is the nature of self-awareness. In this chapter we first discuss how conscious attention for minimal forms of agency may operate in emotions and memory, and then discuss more robust forms of self-awareness, including experiences of flow and meta-awareness. We conclude the chapter with some remarks about consciousness in dreams and in long-term planning, in the light of empirical findings.

4.1 Object-Based Conscious Attention

Since conscious attention requires global access and richly integrated contents, even according to accounts that reject the notion of phenomenality, conscious attention would seem to demand a level of cognitive integration that far exceeds the mere selection of information. Cases of conscious attention, therefore, must include contents that at the very minimum are available for thought, report, action, and memory—contents that are also crossmodally integrated with other information. Another distinctive feature of conscious attention is that its contents are susceptible to semantic evaluation. For these reasons, the type of attention that is necessary for consciousness cannot occur at a subpersonal level or be associated exclusively with procedural memory or motor control. Cases of semantic specificity and ambiguity that have richly integrated contents certainly qualify as cases of conscious attention, as we argued in chapter 1. We mentioned the Necker cube and the duck-rabbit image (see figure 1.2) as examples of such cases. These are ambiguous images that are semantically interpreted and have a rich phenomenology.

In the case of the Necker cube, the image is experienced as "flipping" from one orientation to the other without generating the experience of motion or other changes in the image, other than how it is being *interpreted*. Although the image continues to be interpreted as a cube, its orientation alternates, creating a contrast not only with respect to how it is experienced but also with respect to which aspects of the cube are in focus. With the duck-rabbit image, on the other hand, what changes is its semantic interpretation, and no other changes to the image are experienced, including its orientation. A semantic kind of ambiguity, rather than a strictly perceptual one based on shape, characterizes the conscious and attentional changes one experiences when looking at the duck-rabbit image. These are not images one encounters on a daily basis; indeed, the Necker cube is a purely fictive figure that can be perceived only illusorily. But, given their semantic evaluation, both images seem to require conscious attention. This involves allocating attention to features that give epistemic information about the images and about the way in which they are experienced.[1]

The kinds of changes in attention and experience that are characteristic of ambiguous images, and also occur in cases of binocular rivalry, seem to require conscious attention. This requisite may be extended to all cases of *expertise,* in which the richness in semantic information and how one experiences it exemplify semantic specificity, rather than ambiguity. If one sees a tree as a *Dicksonia antarctica* fern tree, rather than merely as a tree, one attends to it in a different way, accessing epistemic information that produces what is plausibly a different experience—that is, visually experiencing it as a *Dicksonia antarctica* fern tree rather than as a generic tree. This is the basis of the 'content view' of phenomenal contrasts (see Siegel 2010). While we do not endorse or defend this view, we will use the kind of cases it identifies as changes in experienced content to illustrate different forms of perceptual conscious attention.

Based on these considerations concerning expertise, we believe that any kind of process susceptible to cognitive penetrability must be the result of conscious attention, perhaps of the richest possible kind. Cognitive penetrability is a controversial topic: while there is evidence suggesting that some perceptual processes are cognitively penetrable, there seems to be at least as much evidence against it (see discussion in Pylyshyn 1999). Moreover, because our background beliefs seem to have little impact on how we perceive the basic constancies of perceptual scenes, it is highly unlikely that beliefs and concepts have a deep structural impact on these constancies. So here we shall not focus on these "pervasive" forms of penetrability, but will discuss less controversial cases.

Although training and expertise do not seem to alter the basic constancies of how visual objects are structured (e.g., their shape, color, and texture), those factors do have an impact on how attention is *allocated*, which leads to important repercussions with respect to how objects are interpreted and experienced. In addition to the changes described in the Necker cube example, focusing attention on an object can also increase its contrast and apparent size (e.g., see Carrasco 2009). The focusing of attention produces changes in experience that do not seem to be reducible to the features that are being represented, or to changes in the objects that are being perceived; this is the aspect of attention that Block (2010) uses to exemplify his concept of 'mental paint.' Voluntary shifts in attention are important for parsing relevant features of the visual scene and are fundamental for habitual sophisticated behavior, learning, and expertise. These shifts are examples of conscious attention.

In human vision, moreover, a constant interaction between attention and conceptual interpretation produces experiences with highly specific content. We will focus on examples of visualization capacity and semantic recognition that affect dramatically the way visual stimuli are experienced, and will use the distinction between iconic content and conceptual content to illustrate how conscious attention operates. We do not argue in favor of such a distinction; rather, we use it solely to highlight the different cognitive processes involved in consciously attending to the semantic features of a scene. The idea is that even if iconic and conceptual are not different content types, they can be understood as different cognitive capacities, one concerning nonepistemic seeing and the other concerning epistemic seeing—as when one sees something *as* a shape versus seeing it *as* a cube.

A perceptual icon, for the purposes of our discussion, is a stimulus that can be interpreted in multiple, even incompatible, ways without alterations in its basic features. In some cases, like the duck-rabbit image, the changes experienced by the viewer are exclusively dependent on how the stimulus is conceptualized, while in other cases, such as the Necker cube, the changes depend on features that are perceived differently depending on orientation or on which features are prioritized by attention. Neither of these shifts in perception entail changes in the object itself. The corresponding images, with their static and objective features, would constitute the icon, while the experienced changes would be characterized in terms of semantic content, attention, and epistemic access to mutually incompatible interpretations of the stimuli. This shift in semantic focus, we believe, is a distinctive feature of conscious attention.

Besides these cases that involve ambiguity, the interface between semantic information and perceptual icons produces fine-grained representations that play an important role in learning and expertise. One learns where to look and allocate attention when one wants to see a three-dimensional image that can be produced by a random-dot display, or, for example, the 3D stereogram "Magic Eyes" pictures that were popular in the 1990s. Similarly, in a sport like basketball or tennis one learns where to look in order to predict quickly the next move of an opposing player. Other conscious attentional processes are more related to semantic information and epistemic seeing, rather than action and Gestalt effects. Seeing a graph as a function or as the trajectory of a particle, or seeing an animal specifically as a horned owl, depends not only on allocating attention in the right way but also on knowing why the token icon is an instance of a particular conceptualization or semantic category. Conscious attention is necessary to produce these experiential changes that depend on epistemic seeing, or what is called *seeing as*.

The iconicity of perception has played an important theoretical role in theories of visual attention. For instance, perceptual icons are described as the locus in which attention moves, and some authors have described how items of a visual scene can be individuated without being categorically specified (see Pylyshyn 2001). According to some epistemological views, basic beliefs do not need justification from other beliefs because they are directly justified by what is perceptually "given." Fodor (2008), for instance, has argued that discursive-conceptual content is a cognitive refinement of nonconceptual-iconic content. According to views that make this distinction between iconic and epistemic-semantic aspects of stimuli, the iconic aspects may be attended to preconsciously, even preattentively, in a strictly structural sense, according to which the iconic features parse the scene into objects and empty space. So, according to such a distinction, conscious attention operates distinctively or even exclusively at the level of epistemic vision. As mentioned, we are not going to defend this distinction, but will focus on the epistemic cases because of their importance in characterizing conscious attention for objects.

It seems that one can define icons as preinterpreted, strictly structural features of a visual scene. A substantial philosophical and psychological literature on icons and pictorial representations invokes their unique preinterpreted semantics. For instance, the most basic features of pictures and maps are interpreted in terms of a semantics for resemblance and similarity judgments, understood in either metric (location and feature based) characteristics or pre-epistemic inferential notions.[2] While it is controversial to draw a sharp boundary between seeing and epistemic seeing in visual perception, such a distinction helps elucidate the unique features of conscious attention.

The illusion of three-dimensionality seen in random-dot stereograms exemplifies how attention enhances aspects of a scene that are part of phenomenally conscious visual experience, thereby making possible the integration of the image that emerges out of the display.[3] In order to produce this 3D illusion, slightly displaced pairs of dots in a two-dimensional display must be registered by the left and right eye in a way that mimics binocular disparity in natural depth perception. The displays may include thousands of dots, and therefore to compute the displacements perceptually according to categories that distinguish item by item—in this case dot by dot—would be counterproductive as well as cumbersome, because the illusion would thus depend on an effortful epistemic process, counting, and the result in that case would lack the spontaneity that is characteristic of an illusion. It would seem that, in order to generate the illusion of three-dimensionality, the icon must have preconceptual features that can be attended to and pre-epistemically interpreted without cognitive-conceptual effort.

This characterization of icons as pre-epistemic, even preconceptual and preattentive, by which authors mean consciously preattentive, plays an important role in some theories of vision. Austen Clark (2000) and Fodor (2008) propose that unstructured or iconic contents are best characterized as representations that lack principles of categorical individuation. Both authors suggest that these contents nonetheless have *semantic* features. For instance, based on the work of P. F. Strawson concerning feature-placing languages, Clark says: "Feature-placing sentences do not introduce particulars because they lack the wherewithal to distinguish between 'one' and 'another one' or between 'the same one again' and 'a different one.' … The features attributed to places are not properties. Two instances of the same feature are at best qualitatively identical—there is nothing that can distinguish them. … And, in any case, they cannot count to two—or to any number" (Clark 2000, 149).[4]

One way to rationalize this distinction between nonepistemic or "nonindividuating" seeing and more epistemically robust forms of seeing (which require items that are semantically characterized—the kind of item on which conscious attention seems to depend) is by contrasting the act of locating a feature in a perceptual map to that of attending to stimuli of the visual scene as specific types of objects. For instance, whether iconic-region *R* (a subsection consisting of only a red patch) looks like an apple or a tomato does not seem to depend in any straightforward way on the qualitative aspects of such a region. Rather, seeing stimuli as specific objects depends on differentiation based on categorical criteria, which seems to entail epistemic seeing. An apple can be distinguished from other stimuli by virtue of these criteria. Crucially, one attends to apples only by categorizing a region of visual space as an *instance* of the conceptual type 'apple,' rather than by merely focusing on a region of visual space as such. Metrically designated features at regions offer no scaffold to support structural analogies with such specific categorical criteria, even if they are necessary to structure the visual scene. The identification of a location through coordinates does not support categorical discriminations with respect to conceptual types, such as 'apple' or 'grapefruit.' A fair amount of cognitive integration, based on epistemically sensitive information, is required to transform features at locations into an object of conscious attention.

Moreover, as Clark argues, the sensorial specification of regions (place-times) depends upon approximation (Clark 2000, 173–4). A particular sense modality uses coordinates to locate features, such as [F, at coordinates x, y, z, t]. These coordinates have to approximate regions of perceptual space to a certain degree of accuracy. Sensorial discriminations may use different coordinate systems to achieve this goal (e.g., polar or Cartesian coordinates). This kind of coordinate-based approximation, however, need not map with the kind of categorical approximation required to provide a scaffold for semantic structural analogies between icons and discursive symbols. This suggests that feature binding is not systematically associated with conscious attention, and that at best it is associated exclusively with forms of unconscious attention, or alternatively, with forms of non–phenomenally conscious attention.

So far we have assumed that icons require mappings and cognitive integration, suggesting that they are representations. This mapping inherently entails a dissociation between conscious and unconscious attention, leaving open the question of whether attention of a pre-epistemic kind is a necessary condition for conscious attention. We shall now describe an even more minimal sense of 'feature integration,' according to which selective attention might operate on a subpersonal or exclusively sense-specific basis, entailing a larger repertoire of forms of attention that are dissociated from conscious attention. One way of understanding this proposal is that there is a formatting process that transforms unstructured, nonintegrated, cognitively iconic information into a format for global broadcast. According to some versions of higher-order theories, the ideal candidate for such formatting is language, or at least something in a language-like format. Thus, an interpretation of this view is that there is an interface between icons and symbolic representations in a language-like format—a type of program that determines how icons that lack symbolic structure are computed and interpreted as individual symbols, presumably in a language of thought.

As we explained in the previous chapter, although cognitive integration is necessary for access consciousness, it is not sufficient; only semantically rich epistemic access to contents suffices for access conscious attention. In fact, we argued that 'access conscious' attention is best understood in terms of epistemic capacities. Therefore the attentional contents associated with subpersonal processing will not be epistemically accessible and available for global broadcast, as some theories of early attention assume. Consequently, these forms of attention constitute cases of selective unconscious attention, strictly devoted to a single sense modality.

Navigational abilities in animals represent a different case that demands a significant amount of cognitive integration, such as mappings between different metrically structured representations. Some philosophers (e.g., Clark 2000, Haugeland 1998, Millikan 2004, Sloman 1978) have postulated something like iconic representation, based on the distinction between geometrically (or iconically) and logically (or discursively) structured content. John Haugeland (1998) explicitly uses the word 'iconic.' José Bermúdez (1995) and Christopher Peacocke (2001) suggest that some iconic contents have a minimal compositional structure, which they refer to as 'proto-propositional' or 'proto-logical' content. Certainly, one can attend to these geometric features, at least in terms of selective attention. But such content would not be as epistemically relevant as conceptual or logically structured content because iconic information is metrically structured rather than semantically structured. Thus, it seems this proto-semantic characterization of icons could also exemplify a dissociation between consciousness and attention. Because of these considerations, even highly integrated mappings with specific rules for compositionality lack the quality of epistemic access characteristic of conscious attention.

Typical examples of conscious awareness concern objects that one may encounter in daily life. Red apples, the taste of vegemite, and the smell of a rose all illustrate the rich epistemic and experiential content of object perception. Conscious attention requires epistemic access to information and the kind of semantic content that is capable of being combined with other information. The information may concern vivid emotions and properties that are not object based, things like pains, anxiety, or longing. But conscious attention also seems to require objects with highly specific semantic information.

Part of the epistemic and informational complexity of conscious attention concerns the cognitively penetrable aspects of the characterization of objects in relation to the conscious inclinations and urges we have with respect to them. If one visualizes a rectangle "in one's mind" and focuses one's attention at the intersection of two descending lines going from each of the top corners to the opposite lower corners, one realizes that the focus of attention is on the point exactly at the center of the rectangle, where there would be a crossing of those descending diagonal lines. As the research by Stephen Kosslyn (1994) suggests, our capacities for visualization have deep structural and experiential continuities with conscious vision.[5]

Visualization is deeply related to the imagination and to capacities for creativity and literary narrative. According to Elaine Scarry (2001), visualization is fundamental in explaining our capacity to recreate aesthetic and emotional content from simply reading a text that has, perceptually speaking, none of these phenomenal properties. For these reasons, visualization is also a powerful source of emotional and sensual interpretations, which are possible only because of the rich epistemic access and complex phenomenal integration of conscious attention.

The capacity for visualization also includes exclusively epistemic capacities, as when one visualizes the axioms of geometry or learns how to "prove" geometric principles by seeing that they are true of certain visual images. For instance, one can visualize that the axiom of parallels is true by imagining two straight lines extending into "infinity" in both directions without intersecting each other. Many of the visualization cases studied by Kosslyn concern this kind of epistemic process, which is distinctive of conscious attention.[6] The most remarkable kind of visualization, however, is the one noted above concerning emotions and aesthetic information, because it shows that conscious attention ranges over epistemic and empathic information, regardless of whether it concerns objects or not.

4.2 Emotions and Memory

Human consciousness is at its peak in emotional experiences. Aesthetic and moral capacities seem to depend entirely on these experiences. While many emotional experiences happen automatically and without much cognitive effort, it is clear that a fair amount of cognitive enrichment and training are involved in the development of ever more integrated and complex capacities for aesthetic and moral judgments. Art and morality are interwoven in many ways, but we shall focus on the relation between emotions and moral and aesthetic values in the forms of memory that require conscious experience. Our aim is to highlight the cognitive enrichments of conscious attention that are required for memory and emotions. We argue that just as one can distinguish between access and phenomenal consciousness, or between conjunctive selection and the unity of consciousness, one can also distinguish between strictly epistemic and phenomenal memory traces.[7]

We assume that temporal epistemic traces of the sort associated with access consciousness represent events from the past, in compliance with the metric constraints that are required to calculate duration. These representations need not be isomorphic to what they represent, or even be fully detailed vignettes of past events. Rather, these traces need only provide enough information to satisfy the contextually specific epistemic purposes for which they are retrieved. Epistemic traces can frame a background that allows many forms of attention to lead to successful action. Epistemic traces are used to satisfy immediate goals, such as recalling when an event took place in order to decide whether or not another action should now take place. The interaction between semantic and episodic memory produces a rich variety of information sources that guide subjects in their daily activities.

Traces that comply with metric constraints constitute episodic memory—a form of memory that has been identified in many species, and which represents a specific event or experience in time. Declarative memory complies with other epistemic constraints, such as those concerning assertion and testimony. The underlying assumption is that 'remember' is a success term in the case of epistemic traces: subjects who cannot recall episodic or semantic information accurately cannot *remember* this information, and they are either confabulating or misremembering. A particularly problematic aspect of personal narrative memory is that epistemic traces for autobiographical memory rely on a continuously represented subject (Schechtman 1994), which in turn seems to depend fundamentally upon a *causal* connection between the source of information and the memory. Autobiographical memory, besides requiring causality, seems also to require *phenomenal consciousness*.

Epistemic traces must comply with the normativity of knowledge. If one remembers something with the help of an external source such as prompting from a friend with better memory, the trace is not one's own memory. The causal links between the memory trace, the event, and the capacity to retrieve the trace are fundamental to explaining why 'remember' is a success term. The epistemic constraints, therefore, guarantee *knowledge* of reliable temporal traces, which requires the involvement of *capacities* to succeed in identifying and interpreting them. We will expand on these constraints and their consequences for the dissociation between attention and phenomenal consciousness, as well as the overlap between them, in the sections that follow.

4.2.1 Epistemic Traces and Causation

Perceiving duration is different from remembering an event. Although duration can be represented by the phases of a cycle, traces determine a causal connection between events that precede one another, which is incompatible with a cyclical view of time. That is, memory traces involve distinctly individuated events and a causal relation between them. This causal constraint on traces imposes the asymmetry of time, the fixed and closed nature of the past and the openness of the future, as well as the notion that traces represent things on a temporal continuum: a spatiotemporal trajectory that is associated with the self (see Campbell 1994, 110 and chapter 2, and also Hoerl 1999).

According to John Campbell (1994), causality gives us a *conception* of the past. Actually, the fact that traces are causally constrained means that insofar as a subject has the capacity to remember the past and uses that capacity in a relatively robust way by relying on it to form beliefs about previous events, such a subject is a realist about the past—someone who would not consider certain improbable interpretations as epistemically valid. For example, Bertrand Russell (1921) stated that "There is no logical impossibility in the hypothesis that the world sprang into being five minutes ago, exactly as it then was, with a population that 'remembered' a wholly unreal past" (Russell 1921, 104). This is indeed logically possible, but it is also a possibility that no epistemically responsible person would seriously consider. What we would like to highlight about this passage is that Russell assumes that the world would be *phenomenologically* the same and indistinguishable from our current experience had the world sprung into existence five minutes ago.

There are important lessons to draw from this idea with respect to narrative and autobiographical memory, which we address below. For now, we suggest that Ludwig Wittgenstein voices the typical attitude that subjects with standard memory capacities have about the past, an *epistemic* conception of the past: "I believe that I have forebears, and that every human being has them. I believe that there are various cities, and, quite generally, in the main fact of geography and history. I believe that the earth is a body on whose surface we move and that it no more suddenly disappears or the like than any other solid body: this table, this house, this tree, etc. If I wanted to doubt the existence of the earth long before my birth, I should have to doubt all sorts of things that stand fast for me" (Wittgenstein 1969,

31). This passage presents a broadly Moorean endorsement of the indubitable epistemic status of commonsensical beliefs about the past. Even if it is possible that the past is illusory, one *must* be a realist about the past. Otherwise, very few things about the world make sense. The asymmetry of time and the causal constraint on traces are fundamental to justify the beliefs that Wittgenstein mentions. Our focus here, of course, is exclusively on the cognitive aspects of traces, rather than their metaphysical grounding.

How is the causal origin of long-term and short-term mnemonic traces epistemically secured? *Reality monitoring* is the epistemic activity that verifies the origin of information, thereby distinguishing different informational sources. We explained why this capacity seems to be intimately associated with mechanisms for monitoring reliable information and why it is indispensable for access consciousness and epistemic agency. Reality monitoring for traces seeks to confirm that such traces are causally linked to events that one *perceived*, rather than those that one merely thought about or dreamed (Johnson 1991). This parsing of traces according to their causal origin is crucial to satisfy the epistemic constraints. Many factors have to be in place for epistemic tracing. Mnemonic traces need to satisfy very demanding requirements, besides the causal and metric constraints for accurate duration and temporal order judgments. Because of the importance of reality monitoring, these requirements include the reliable *storage* and *formatting* of traces, which depends on a great deal of cognitive integration. *Retrieval*, too, must be epistemically reliable.

4.2.2 Epistemic Temporal Binding and Cognitive Integration: Storage and Epistemic Tracing

The view that mnemonic traces are literally "stored," as if they were deposited in neatly designated cabinets somewhere in our brain, is implausible for several reasons. For example, adequate storage does not seem to require huge amounts of "space" to keep individual traces. Ideally, the storage of traces must include information about the relation between specific facts or events stored in different traces. Storage may be map-like, or isomorphic to what it represents, but it is unlikely that the degree of resemblance has to be very high or that storage requires atomically individuated traces, tagged centrally by a storage and retrieval system. The alternative to itemized storage is a distributed network model of encoding, which is also problematic. It is true that a holistic and distributed format for memory storage, realized by a connectionist network, is more realistic than itemized storage; but that too might have some epistemic shortcomings, particularly with respect to reality monitoring.

An advantage of a holistic account is that it allows for a dispositional approach to memory capacities: traces dispose the subject to retrieve information in specific ways, which are determined by the roles the traces play in the network (Rumelhart et al. 1986). The flexibility of this account, however, comes at an epistemic price. Too much mixture of information may violate the causal and metric constraints, which require informational specificity regarding sources of information, rather than mere functional roles within a framework. Because traces are recreated in this framework, rather than retrieved from storage, reality monitoring becomes problematic. A possible solution is to propose supervision and central monitoring; but this jeopardizes the flexibility, context sensitivity, and parsimony of the approach. In particular, such centralized monitoring for any type of memory retrieval seems to require an unacceptably demanding kind of voluntary effort and attention. While attention should play a role in reality monitoring, requirements of voluntary effort and centralized computations seem excessive.

Based on these considerations, it seems that any account of epistemically constrained storage must find a balance between itemized storage and a fully holistic regeneration of traces. A proposal for solving this problem is to increase the *externality* of the cues that trigger the retrieval of a trace, and perhaps even the externality of stored information, as opposed to internal cues that require monitoring and a very high degree of cognitive integration among traces. This interaction between context, external information, and trace memory is likely to be at the center of a plausible account of epistemically constrained storage, and it necessitates the functions of attention. While it need not be conscious attention of the phenomenal kind, attention certainly plays a role in reality monitoring and can thus be labeled 'access conscious attention.'

Another trade-off concerns the degree of filtering and modification that context should exercise on traces. Emotional and social aspects of a trace may modify how the trace is stored and interpreted, thereby changing or even eliminating causal and metric features of the trace. The trade-off is that storage must not modify the trace to such a degree that it loses its epistemic characteristics, but it must also not preserve it so rigidly that it cannot be interpreted in different ways. There is an analogy here between conscious visual attention and conscious attention for memories: *conscious attention enriches the variety of possible interpretations one can have of a single object or an event that is remembered*. We expand on this issue below. Notice that this trade-off concerns the *influence* of emotional and social biases, rather than their atomistic or holistic format.

Finally, there is a trade-off that limits the temporal range of a trace. It would be ideal to have immediate epistemic access to as much reliable information as possible, based on limitless storage of detail and temporal span. For responsible epistemic agency, however, it suffices that one succeeds in storing the most useful and reliable information needed for concrete epistemic tasks, such that one is always equipped to specify the causal origin of traces. It would be useless to store information about every single event that occurs during the day. The brain does not store every detail about an event for long periods of time. There are cases of people with remarkable memory storage, but they are the exception to the rule.

Many traces are normally stored for only a brief time, independently of the limits of memory capacity. Short-term memory presents a bottleneck with respect to how much information one can hold at any time, with crucial repercussions on sense-specific and crossmodal attention. The interaction between short-term and long-term memory is an important area of study, since it helps us understand how things become encoded into long-term memory for later retrieval. The trade-off about range, however, does not concern bottlenecks or limits of short-term memory capacity. To illustrate this trade-off, think of familiar examples in which it is normal and quite safe, epistemically speaking, to forget something. You can, for instance, remember what you had for breakfast today, but not what you had for breakfast three years ago. The normative formulation is also valid: I should remember what I had for breakfast today, but I should not be expected to remember what I had for breakfast three years ago. Moreover, we must not keep useless information for long periods, and at the same time we must not lose traces after a few hours or days. It would be functionally challenging if no crucial information could be kept for relatively long periods: autobiographical narratives, based on epistemically constrained memories, could not be formed.

These trade-offs concerning format, biases that impede accurate storage, and temporal range must be calibrated to achieve ideal epistemic performance, and they all involve attention in one way or another. They do not, however, seem to necessitate phenomenal experiences. Rather, they all seem to require detailed access to the contents of long-term memory.

4.2.3 Epistemic Retrieval and Attention to Detail

Reliable memory retrieval is epistemic access to traces. Traces can be evidence for the justification of a belief only if they comply with causality and metric constraints. As mentioned, epistemically constrained memory for time corresponds to the memory system that Endel Tulving (2002) called 'episodic memory,' which allows subjects to register and retrieve information about particular events according to their time of occurrence. As we explain below, the relationship between this system and the autobiographical memory system is more problematic than is generally assumed, which has important repercussions for the theoretical characterization of conscious attention. Retrieval, like any other case of epistemic access, has as a central consideration the *quality* of such access.

There is another trade-off with respect to the quality of access to traces, which we call *Funes's principle*, based on the short story by J. L. Borges: remembering in excruciating detail is at one extreme of a continuum and remembering in the most abstract and ambiguous way is at the other. The brain does a good job of striking a balance between the two extremes.[8] The access we normally have to information is never so rich as to allow judgments concerning the specific color, location, and identity of every object we encountered exactly three years ago, or even a week ago. But we do remember enough to think about what we did three years ago, roughly speaking, or what objects we may have encountered a week ago. Detailed memory is obviously crucial in enrichments to conscious attention. Identifying a place, recognizing objects, and relating past events to current information all require the interaction of at least access conscious attention and memory—and perhaps, in some cases, phenomenal consciousness. In fact, attention and access to content seem to be crucial for modulating how best to achieve this equilibrium concerning detail.

Besides the equilibrium point associated with Funes's principle, there is another equilibrium point that needs to be reached between *suppression and intrusion*. While Funes's principle concerns detail and particularity, the suppression and intrusion balance concerns the recurrence of memories and our capacity to retrieve them at will. Although this continuum also applies to attending to (or ignoring) thoughts, unwanted recurrent memories are produced by specific mechanisms that have unique cognitive implications. Memories that come to mind spontaneously may be useful, for example when you remember an important task you have been forgetting. But in many cases, such memories enter into our conscious awareness in a strange and troubling way. Too much intrusion is extremely disruptive. In the case of too much suppression, however, memories would not be generated spontaneously and one would have trouble linking memories and contexts, which would preclude flexible interactions between traces and the environment. Excessive suppression could increase the rigidity of traces, which then could be retrieved only under considerable voluntary effort, placing unnecessary demands on *conscious voluntary attention*.

The mechanisms for intrusion and suppression have critical implications for the stability of our personal narratives, and they operate in intriguing ways. For example, depression has been associated with intrusion, but not specifically with suppression (Schmidt et al. 2009). Suppression has unconscious components to almost the same degree as intrusion. Conscious effort to suppress a thought (e.g., "don't think of a white bear") has been found to generate the opposite effect, *exacerbating* the intrusion of such thought (Wegner 1994). Daniel Wegner proposed the *ironic processing* theory in order to explain this phenomenon. The main idea is that resources for suppression ironically highlight the suppressed thought for targeted recurrent-attention processes, producing intrusion.

Suppression features centrally in Sigmund Freud's theory of memory. Repressed memories produce conflicts that the individual needs to overcome by revisiting her own narrative, modifying it so that the suppressed memories are associated with current experiences in relevant and insightful, even liberating, ways. Attention is crucially involved in this process. Although Freud's theory of memory is insufficient to capture the complexity of mnemonic processes, suppression is certainly an activity that the brain has to balance. It is a familiar experience to suddenly remember an event only after *recognizing* a place, for example, or the face of a friend. The trace was stored, but the retrieval process was not epistemically facilitated by internal cues, so retrieval of the trace required attention to external cues. Many forms of unconscious suppression are eliminated roughly in this way: *the right cue must be attended to* in order for the trace to be epistemically accessible.

If such external cuing were needed for access to most traces, however, we would be epistemically handicapped and fully dependent on the contingencies of the environment for successful retrieval. This would be the case if suppression were prevalent, even if one had epistemically impeccable memory storage. The positive side of suppression, both conscious and unconscious, is that it prevents rumination and constant reevaluation of what one could have done in the past. Thus, a consequence of adequate suppression is the prevention of obsessive routines and behaviors.

The importance of the balance between intrusion and suppression is evident in conditions where one or the other dominates, with deeply damaging consequences for the subject. Recurrent thoughts and memories can be disruptive and tormenting. Intrusion in post-traumatic stress disorder (PTSD) is a paradigmatic case of epistemically inadequate retrieval with powerful self-narrative effects. Intentionally and nonintentionally reexperienced trauma memories are incredibly vivid, and researchers argue that they should not be called intrusive *thoughts* because they are far more damaging to autobiographical narratives than intrusive thoughts. Intrusive memories, therefore, should be carefully distinguished from the intrusive thoughts associated with short-term suppression tasks and rumination (see Ehlers, Hackmann, and Michael 2004, and references therein). Patients very rarely describe these memories as intrusive thoughts, and they are so vivid that some patients lose completely their *temporal perspective*:

> The intrusive re-experiencing symptoms in PTSD appear to lack one of the defining features of episodic memories, the awareness that the content of the memory is *something from the past*. In a dissociative flashback the individual loses all awareness of present surroundings, and literally appears to relive the experience. The sensory impressions are re-experienced as if they were features of something happening right now, rather than being aspects of memories from the past. Also, the emotions (including physical reactions and motor responses) accompanying them are the same as those experienced at the time. (Ehlers, Hackmann, and Michael 2004, 404)

The intrusive trauma memory is so powerful that it overwhelms the reality monitoring system, which takes information from memory to be currently experienced information. Precisely because reality monitoring fails, attention to a past episode (including the vivid sensations and emotions associated with it) is experienced as attention to a present event. Notice that the information is not entirely *false*. On the contrary, most of it is painfully true. The epistemic constraint that is absent is not truth-tracking retrieval, but *temporal* information: the emotional reactions are not registered as responses to information that occurred in the past. Less dramatic cases involve attended cues that trigger emotions and reactions that are associated with the traumatic event but without any *recollection* of the event (Ehlers, Hackmann, and Michael 2004, 405). In the one case the trauma narrative is relived without any temporal perspective, while in the other it evokes reactions without being explicitly recalled.

Processes of intrusion and suppression may also be altered, conditioned, or enhanced by interactions with subjects that have a very high quality epistemic access to events that one experienced with them. This collaborative interaction between storage, intrusion, and suppression across subjects could be categorized as *transactive memory* processes, a term introduced by Wegner (1987). The collective memory of groups or organizations, families, and close friends is characterized by this transactive interpersonal cuing, both for specific memory retrieval and for the reshaping and preservation of narratives. It also plays an important epistemic role, as a kind of *collective* reality monitoring.

Thus, a memory system has to provide optimal solutions for a variety of trade-offs concerning storage, format, and retrieval. It may seem that the most important function of the memory system is to provide accurate information about the past. Storage and retrieval must be accurate and reliable, be causally linked to information, and produce more truth than falsehood. Notwithstanding the complexity of negotiating such trade-offs, this reality monitoring approach to memory, based on access and attention, takes 'remembering' to be a success term that describes an epistemic achievement. Access consciousness seems to best characterize these achievements.

Epistemic constraints, as this section shows, are intricate and present many difficulties. Yet, we shall argue that these problems—on which most accounts of mnemonic traces have focused—concern only one type of mnemonic cognitive integration. The view that there is a different type of memory integration, which we shall call *phenomenalism*, has its own problems. Although the epistemic view seems to be true about episodic and semantic memory, autobiographical memory requires a balance between epistemic achievements and phenomenalism. Accordingly, the distinction between epistemic and phenomenal traces is analogous to the distinction between access and phenomenal conscious attention. In terms of structure, this distinction also resonates with the difference between conjunctive and subsumptive conscious attention.

4.2.4 Phenomenal Traces

While memory can be defined as a truth-seeking mechanism, phenomenalism conceives of memory as a constructive and integrative process. The former sees it as a criminal investigator, while the latter sees it as a theater director. Research has shown that memory is indeed malleable and susceptible to manipulation (Loftus 2005). This does not mean that memory is generally unreliable, or that we should be skeptics about the past. On the contrary, we could not do many basic actions without having accurate memories. The malleability of traces, however, presents the challenge that memory does not have the exclusive function of providing accurate information about the past; an equally important function is to *make sense of the past*.

According to phenomenalism, what is most intriguing about memory is what it evokes, rather than the accuracy level of the information contained in the memory. Obviously, it is important to retrieve accurate memories, but what is distinctive about how we *consciously* remember the past is that it makes us understand, feel, and relive events with powerful effects on imagery and the imagination. For the phenomenalist, although all memories of dreams are false because they are about unreal events, they are as distinctively narrative as normal wakeful memories and thus should be studied *as memories*.

Memory of dreams is unique. Everything from storage to retrieval is different about our memory of dreams compared to our waking memories. In spite of the vivacity of dreams, there is active suppression during storage and retrieval—a constant form of reality monitoring. But dream memories seem to be made, as it were, of the same material as waking ones, and we seem to have the same capacities for conscious attention, although, of course, here conscious attention of the phenomenal kind is *running free*, as it does in hallucinations.

Some of the memory processes during sleep and dreaming are part of the epistemically constrained memory system. The consolidation of epistemic traces, for instance, varies between wake and sleep periods. Sleep seems to increase memory consolidation—at least of the semantic or declarative type (Payne et al. 2012). Consolidation is necessary for successful future voluntary retrieval; therefore, unconscious sleep-consolidation plays an important *epistemic* role. Although dreams may be related to learning and creativity, their most important characteristics are not epistemic. Descartes famously used the powerful and convincing experiences we have in dreams to raise epistemic *problems*, which some philosophers think are unanswerable. Instead of focusing on skepticism, we shall present a positive characterization of the unique qualities of dream awareness, with an emphasis on the commonalities between dreams and other phenomenal mnemonic traces—commonalities that are captured in the claim that dreams, like hallucinations, are introspectively indistinguishable from veridical experiences. Thus, although phenomenal traces can be epistemically constrained by reality monitoring, they need not be.

Epistemic traces are schematic: they contain a few entries, concerning location, time of the event, and perhaps most important, verification that the information was perceived rather than daydreamed or dreamed. The vivacity, the emotional intensity, and the perspective-dependent features of the trace obey principles for *phenomenal integration*, which are not subject to metric or causal-cognitive constraints. This distinction resonates with the distinction between access conscious integration and phenomenal unity discussed in the previous chapter: like subsumption, phenomenal mnemonic integration does not seem to have a strictly epistemic role.

With autobiographical memory, perspective includes not only the point of view of the subject but also her narrative. The same epistemic trace could, therefore, be experienced with different levels of absorption, empathy, and vividness. The articulation of emotions according to different narratives is yet another dimension of variability for phenomenal traces. The same epistemic trace with invariantly accurate information across many types of phenomenal traces will have different emotional and cognitive effects, depending upon different levels of phenomenal integration.

Perspective illustrates this point. As has been noted, one can remember an event either from the perspective of "the observer" or from a selfless perspective that includes the environment as a whole, such as a room or a dance hall (Rice and Rubin 2009).[9] The same event can be remembered accurately with different narrative effects. To give a concrete example, suppose one remembers being in a serious car accident. The epistemic trace includes duration, location, and basic factual information. The phenomenal trace, however, includes perspective. One could remember the event from "the inside." People report that in car accidents time slows down and that very specific details, like the way the windshield shatters, seem to be incredibly vivid from one's own perspective, *as observer*. Alternatively, one could remember the event from the side of the road, as it were. When recalling the windshield shattering, one would observe the accident, and *oneself*, as well as other passengers, from a specific vantage point, as in "out of body" experiences. The perspectives one can take are varied. One also could remember the event from above, for instance.

Perspective is one way to illustrate these problems, but there are other examples regarding the ornamentation and emotional infusion of phenomenal traces that determine how they are phenomenally unified. We shall describe two of these examples: Proustian flooding and Tolstoy's principle.

Proustian Flooding Smells and sounds evoke powerful phenomenal traces. When these traces are triggered, what is fundamental about them is the vivacity and significance of what they evoke in the subject. They can be accurate, occurring in tandem with epistemic information, but the set of evoked recollections saturate the schematic epistemic trace in colorful ways, opening up a spectrum of *phenomenal variation* with respect to the same memory. Metaphorically speaking, triggering these traces is more like opening a door rather than neatly retrieving specific information about the past. These traces are surprising and produce profound emotional and aesthetic effects. We will call this involuntary process of rich recollection based on external cues 'Proustian flooding.'

In *Remembrance of Things Past*, Marcel Proust described these aspects of phenomenal trace recollection: "And suddenly the memory returns ... when from a long-distant past nothing subsists, after the people are dead, after the things are broken and scattered, still, alone, more fragile, but with more vitality, more unsubstantial, more persistent, more faithful, the smell and taste of things remain poised a long time, like souls, ready to remind us, waiting and hoping for their moment, amid the ruins of all the rest; and bear unfaltering, in the tiny and almost impalpable drop of their essence, the vast structure of recollection" (Proust 1922, 61).

The power of a scent to recall for us an event or a person illustrates Proustian flooding. It has been confirmed empirically that odors are powerful cues for particularly evocative and vivid autobiographical memories.[10] In his work, Proust described how cues can trigger memories automatically in the sense that their *retrieval* requires no control or conscious guidance. Yet these traces produce powerful phenomenological changes. If the memory triggered is very pleasant, a torrent of pleasing experiences may be associated with it—the face of a person, the surroundings, and other sensorial memories. The same can occur with unpleasant memories. Music has a similar effect on the spontaneous retrieval of phenomenal traces. When one listens to a song one has not heard in a long time—a Beatles record, a nursery rhyme—the experience can be one of *immersion* in the past. One feels as if the memories of a whole period of her distant past suddenly rush in.

Proustian flooding is one way in which traces enrich and make our phenomenally conscious lives interesting and surprising, as well as *meaningfully continuous* with our past. Crucially, these traces are not discretely retrieved as independent items in a temporally organized way. Rather, the traces constitute a form of revelation in which many memories are experienced as a single contextual whole. The common feeling of nostalgia, triggered by a perceptual experience—a song, a scent, a photograph—is another instance of the same kind of memory experience. This resurgence of recollections, incidentally, seems to be fundamental for the retrieval of dreams; a single trace may trigger an unexpectedly complex dream narrative. We expand on this phenomenon below.

The mechanism that triggers biographically arranged phenomenal traces is associated with aesthetic experiences and emotions related to morality. Some studies indicate that aesthetic and moral evaluations may depend on the same neural correlates and may have a common evolutionary explanation.[11] Obviously, a bias toward identifying the truth about the past with what is beautiful is epistemically pernicious, but the common origin of moral and aesthetic judgments is compatible with reality monitoring, thus allowing for *phenomenally conscious enrichments* such as Proustian flooding. These enrichments are clearly associated with the new knowledge that Jackson's Mary learns. One could imagine a pre-revelation Mary (before she experiences red for the first time) complaining: "I don't want to always be *right* about what color surfaces are red, I want to *understand* and *empathize* with what others feel. I want to understand, for instance, what is so *evocative* about those Rothko paintings."[12] Creativity and insight are considered aspects of such enrichments. Conscious attention also seems to be necessary to produce this kind of cognitive integration into a phenomenal unity.

It must be emphasized that the traces evoked need not be *chronologically* arranged in order to be suggestively powerful. On the cont what the "flooding" metaphor is supposed to capture. Several " of phenomenal traces rush into consciousness simultaneously, memories "flashed before one's mind." Moreover, the traces need not be entirely accurate to produce reliable autobiographical information. In fact, as long as they are not fully confabulated, their precise accuracy seems irrelevant, because what is distinctive about these traces is their self-narrative significance. The way these memories appear all at once leads to another problematic trade-off, between accurate causal tracing and the evocative character of a set of experiences.

Samuel Beckett captured this trade-off in *Krapp's Last Tape*. Krapp has an archive of tapes that are chronologically organized, and he is looking for one that would bring a significant and powerful memory. As Krapp proceeds, it becomes clear that the evocative power of a set of memories cannot be captured by their accuracy, temporal order, or causal origin *alone*. Krapp's file helps him locate a memory trace registered in a tape and verify its accuracy and temporal order. In principle, Krapp can do this with respect to any set of traces archived in the tapes, but these filed memories are a pile of traces. If they are to evoke the meaning and significance of certain events, the epistemic information about temporal order and causality must be infused with autobiographical narrative. Proustian flooding helps perform this infusion, and it also helps explain how attention is automatically grabbed by the evocative and meaningful experiences—not because of their accuracy but because of their significance.

Proustian flooding is not the mechanism of abnormal memories such as PTSD memories. Proustian memories are not recurrent or traumatic, and one does not lose one's temporal perspective when they occur. Reality monitoring works perfectly well amid Proustian flooding, and the way in which perceptual cues trigger these traces is a fundamental characteristic of *phenomenally conscious memory*, which seems to depend on conscious attention. In fact, we take this to be an illustration of how attention and phenomenal integration can occur in tandem. Although too much evocative power may lead to epistemic problems like confabulation, the contexts in which this would be problematic seem to be limited to cases that demand greater accuracy than is usually expected, such as in a court of law. Proustian integration articulates emotions and aesthetic experiences, and it involves spontaneous retrieval. The next form of phenomenal tracing concerns voluntary retrieval and a purposeful attempt to accurately describe past events based on such retrieval.

Density: Tolstoy's Principle A difficulty regarding phenomenal traces concerns the density and richness of the information contained in self-narrative. The voluntary and effortful reconstruction of autobiographical narratives depends not only on tracing external and causally constrained events, but also on the way in which the associated conscious experiences are integrated in a stretch of *self-conscious* awareness. There is a trade-off between capturing the asymmetric and linear structure of epistemic traces and the complexity of a stream of conscious awareness. The trade-off is that the more one stays within a linear structure, the less accurate the description of the stream of consciousness *as experienced* by the subject—and the more one departs from a linear narrative, the higher the risks of confabulation. We call this trade-off 'Tolstoy's principle' because of Leo Tolstoy's insights concerning the insufficiency of causally constrained events to accurately narrate the past. In contrast to Proustian flooding, the risks and benefits of voluntarily enriching one's narrative are different. In retrieval of Proustian flooding, one is not in command of the reconstructive task, but in voluntary conscious retrieval one is actively describing a stretch of conscious awareness (of any kind, including dreams).

The variation created by the voluntary tracing of consciously remembered events is illustrated by the change in perspective mentioned above in the auto-accident example. We focus on two changes in perspective: subjective and temporal. In describing what one did yesterday or what one dreamed last night, as in any conscious effort to trace autobiographical memories, one may take a first-person or a scenario-field perspective. Perhaps, for example, you took part in an unproductive meeting where people were confrontational. You could recall in detail the tension of the meeting, the facial expressions of the attendees, and what they said or did not say. Or, you could recall the event very generally, focusing instead on what you were thinking at the time. If it was a boring dinner party, you might recall the monotonous conversation and the thoughts you had about the conversations, or alternatively, you might just recall what you were daydreaming about during the party. The tracing of the events is dense because there is no simple and unique way of capturing everything, and what is remembered radically depends on how *attention is allocated to thoughts or external events that are consciously experienced*. This interaction between attention to thoughts and to external events is possible only as a form of conscious attention because it assumes conscious experiences as the focus of attention.

The same occurs with *temporal perspectives*. Just as one can change perspective from a subject-centered narrative to a scenario narrative, one can switch from a causally constrained linear temporal progression to a backward-looking narrative or even a timeless narrative, in order to be more precise and insightful about phenomenal traces. Examples of how the most meaningful of a set of possible narratives moves from the present into the past, meandering and creating pockets of durationless experiential plateaus, abound in literature. Similarly, the ability to describe several possible temporal trajectories, all equally legitimate, backward and forward from a single vantage point, as if they were frozen at the moment of mnemonic tracing, is crucial for self-narrative. Timeless perspectives are useful to describe the intricacy of an event that is fraught with sub-events. This phenomenon of assuming an accurate timeless perspective can be called 'nesting,' because one event is traced at one point in time in all its intricacies, as one narrative falls within another. In all these cases the meaningful association of phenomenal traces is more important than their causal linear structure or their strict epistemic accuracy. Tracing in dreams has the same qualities, an issue we address in section 4.4.

All these interactions between emotion, memory, and attention require conscious attention. That is, they can only be integrated and phenomenally experienced when a conscious attention is directed to them. The role of attention for thoughts and intention, as well as for the higher forms of self-awareness required for autobiographical memory, exemplifies a form of attention that is not only conscious attention, but seems to be more complex because it involves self-aware conscious attention. As mentioned in the previous chapters, it is implausible to think that this form of conscious attention is necessary either for conscious awareness or for attention capacities of any kind. There is no doubt, however, that forms of self-aware conscious attention are at the basis of what we call 'phenomenal tracing' for autobiographical memories, and that they generate some of the richest and most complex manifestations of conscious awareness. Next we discuss aspects of this type of conscious attention.

4.3 Awareness and Self-Awareness: Perceived Effort and Agency

Perceived effort is intimately related to voluntary attention. There are tasks, like solving a mathematical equation, thinking about how to pay bills more efficiently, or avoiding an immediate problem, that require a fair amount of sustained voluntary attention directed toward thoughts, goals, and plans. These seem to be clear cases of self-aware conscious attention. Effortless attention, however, challenges the assumption that all forms of highly integrative conscious attention for demanding tasks require this kind of introspective conscious attention directed toward contemporaneous thoughts and plans. As we mentioned earlier, the experience of 'flow' is an example of effortless attention. For instance, when our friend Lloyd plays the piano beautifully, he does not have a self-reflective attention directed toward the planning of which keys to hit or remembering how to read the notes, even though the task requires great skill; instead, he "loses himself" in the flow of playing. Thus, it seems that one type of conscious attention is almost exclusively directed toward actions that are highly demanding without a focus on the self, while another type of conscious attention targets voluntary effort by focusing on experiences of the self, such as one's thoughts and plans.

As Lau and collaborators (2004) have argued, evidence shows that brain areas related to intention can be differentiated from those associated with action, and that attention can be allocated either to the intention to act or to the action itself. It would seem as if this kind of attention to intention is not necessary for experiences of flow, because what characterizes such experiences is a loss of self. The importance of the contrast between flow-related conscious attention and attention to intention is that they could constitute two different forms of highly integrative conscious attention: *conscious attention for complex tasks,* and *self-conscious attention.*

There are two ways of understanding self-conscious attention. One is in terms of higher-order theories, which do not require awareness of a self in the processes of targeting a lower-level representation for higher-order cognition. The idea is that in the processes of targeting a lower-level representation, that is, an object of sense-specific selective attention, conscious attention does the targeting without requiring an explicit representation of the self. We discussed such theories in the previous chapter, and showed that they entail a dissociation between consciousness and attention. Here we simply recall that they do not require Cartesian self-reflective accounts of conscious attention, despite the fact that conscious attention produces experiences as experienced *by* oneself.

An alternative construal of self-conscious attention is that reflection on the self is constitutive of self-conscious attention, at least with respect to thoughts and intentions. We will explore some options to illustrate this idea, particularly in connection with autobiographical memory, and argue that even in these cases it is not entirely clear that an explicit representation of a self is required, although autobiographical memory does entail that one is attending to oneself in some sense. While we do not deny that there might be forms of reflective and metacognitive conscious attention, our main contention is that one must distinguish the type of conscious attention associated with flow and higher-order theories from the type of conscious attention directed to the intentions and thoughts of an explicitly represented 'self.' This distinction has to be made systematically, and it suggests there might be two forms of conscious attention, likely to be dissociated from one another. One form would be deeply associated with experiences of flow, the other with experiences of perceived effort.

The distinction we are proposing entails that the 'transparency method' used to establish knowledge of one's own beliefs may be problematic. According to this method, all a subject needs in order to *know* that she has a belief with a particular content is to *simply believe* such content. From believing that "the universe is expanding" one can conclude that "I believe that the universe is expanding." Although it has been suggested that the capacity to form beliefs entails that one can conclude without cognitive deliberation or evidence gathering that one is forming such beliefs (see Byrne 2005), it seems that in order to *know*—to truly believe with justification, and to comply with a safety or sensitivity condition—one must explicitly use rational deliberation, including evidence gathering, to be epistemically entitled to assert such knowledge (see Moran 2001).[13] More specifically, it seems that one needs access monitoring, or reality monitoring, to have self-knowledge, because establishing the origin of information seems crucial for knowledge of one's own beliefs. We expand on this in the following section, where we discuss dreams.

4.4 Dreams, Perceptual States, and Planning

Memory traces related to dreams, or oneiric mnemonic traces, are puzzling. Those who consider 'remember' a success term would of course deny that dream traces are authentic memories, suitable for supporting beliefs.

Skepticism, however, does not eliminate the evocative power of these traces. Since antiquity, dream interpretation has been an important spiritual, and even an epistemically productive, activity. Accounts of revelations as well as testimonies of personal insight in dreams are remarkably common.[14] Dream retrieval and interpretation obviously involve memory for dreams, which differs radically from waking memory.

Memory storage in dreams is bizarre and paradoxical. The way dream traces are stored defies any linear or causal order. The "sequence" of a dream generally involves impossible transitions, as from one location to a radically different one, or involving different people at incompatibly different times. For this reason, Maurice Halbwachs (1952) said that it is impossible to remember anything at all in dreams, which includes, according to Halbwachs, *all* forms of memory. Halbwachs also claims that memory in dreams is so "inert and drowsy" that the dreamer cannot remember that a person who had died a long time ago is dead, and so "interacts" with the person as if nothing has happed—an observation that he attributes to Lucretius (Halbwachs 1952, 41). Lucid dreaming and meaningful experiences in dreams show that this cannot be adequate to describe all dream memory, but it is true that dream storage obeys no rules of causality or linearity.

The retrieval of oneiric traces is unique. The window of opportunity to recover a dream narrative is narrow; generally all traces quickly vanish after one is fully awake. The traces are compartmentalized in sections of a dream that succeed each other, or in sequences of what one considers to be different dreams. Recovering a single trace of a dream requires significant effortful conscious attention, and such effortful retrieval attempts are prone to confabulation. Interestingly, recovering a single trace generally leads to the retrieval of substantial portions of the dream, sometimes even the whole dream. Yet, all mnemonic traces from a dream are extremely weak and easily distorted and suppressed, in spite of their powerful emotional content.[15]

Unlike waking phenomenal traces, it seems that dream memories are subject to a systematic bias in which our brain tends toward suppressing them rather than negotiating trade-offs between evocative power and epistemic value. Perhaps this is the case because, as the research by Johnson (1991) suggests, this is the only way to keep reality monitoring functioning reliably. One must go through considerable conscious effort to retrieve dream memories, and even when one succeeds, the trace quickly vanishes, unlike wakeful phenomenal traces. In dream retrieval, conscious attention occurs simultaneously with experienced effort and self-awareness. Strikingly, dream phenomenal traces are not distinctive in their phenomenal qualitative character: one could easily imagine that what one is currently experiencing, including a memory, is a dream, as Descartes famously argued. So the difference must originate exclusively in reality monitoring, which manifests in strong suppression and poor retrieval. Culture could be a factor too. Dreams are now considered odd things we experience while we sleep, and people are rarely encouraged to keep track of their dreams or interpret them, unlike people in earlier cultures.

Yet we remember our dreams, find meaning in them, and hold on to those dreams we find most insightful. Undoubtedly, dreams are infused with surreal confabulation. We may look at a place that resembles a school and recognize somehow that it is our house, or see a person that looks very different from who we know the person to be—a recognition despite absence of resemblance. But with respect to their phenomenal character, dream traces strongly resemble waking ones. Dreams reveal crucial aspects of the perspective of the *observer* that can be contrasted with wakeful forms of memory retrieval. Epistemic constraints are insufficient to understand such a perspective, because the rules for phenomenological integration seem to be the same in dreams and in waking conscious awareness. Certainly, a study of mnemonic traces cannot ignore the distinction between epistemic and phenomenal traces, including oneiric phenomenal traces.

4.4.1 Dreaming and Daydreaming

There are at least two views about dreams. One is that dream experiences are just like waking ones. More precisely, this view proposes that although dreams are epistemically deficient, they are phenomenally identical to wakeful experiences with respect to their qualitative character. We can call this the *continuity thesis*. An alternative view denies the continuity thesis, either by viewing dreams strictly in terms of narrative confabulation, or by explaining dream traces as an odd form of imagery, phenomenally different than perceptual-like experiences.

The empirical evidence supports the continuity thesis.[16] Not only are dream traces similar in phenomenology, they are also similar in cognitive and autobiographical content. Moreover, dreams seem to be continuous with daydreaming, which seems in an even more robust sense to be a wakeful variant of *planning*, since the same areas of the brain are recruited for dreaming and for planning or daydreaming (Fox et al. 2013). The so-called 'stimulus-independent thought' recruits the same area of the brain in dreaming and daydreaming: an area associated with the 'default network.'[17] This is consistent with what we said in the previous section concerning the necessary involvement of conscious attention when the targets are thoughts, action plans, and experiences, regardless of their source.

The integration and consolidation of a personal narrative is a fundamental cognitive function, deeply associated with autobiographical memory. Immersion in the present moment can be overwhelming, and a torrent of sensory impressions can never by itself help create a personal narrative. The ability to do so requires isolating the observer from the torrent, in a *de se* belief fashion distancing *oneself* from it. Phenomenal traces make this possible in daydreaming and dreaming by consolidating experiences independently of duration and cause. For this reason, narrative consolidation is crucial for robust forms of long-term planning and the successful management of long-term goals (Smallwood, Ruby, and Singer 2013). The sources of information that must be assimilated, as with phenomenal consciousness in general, vastly exceed perception, and include emotions and social cognition.

With respect to *perceptual* phenomenal consciousness, dreams and perception share more in common than is generally acknowledged. The bizarreness of dreams seems to concern only the overall structure of their narrative, rather than reflecting specific differences in the phenomenology of perception. One may dream odd things, but people have eyes, "color" is stable, things look three-dimensional, and one can "touch" and "hear" the same textures and sounds. Consequently, as the continuity thesis predicts, the perceptual constancies of waking consciousness are for the most part also present in dreams. In spite of the empirical support for the continuity thesis, however, it faces well-known theoretical objections. One is that a naive version of the continuity thesis seems too strong. Dreams are very different from daydreams and waking experiences in at least two ways. First, while awake one has a stable connection with other types of memories, and second, wakeful content can be quickly updated with epistemically constrained perceptual information. Since this is impossible in dreams, the continuity thesis must be false.

A reply to this objection is found in Descartes's insightful discussion of dreams. The continuity thesis does not require strict resemblance across stretches of conscious awareness, but only a much weaker standard of similarity. One need not prove that every dream could be experienced in waking life and vice versa, obviously an absurd proposal. Instead, all one needs to show is that one *could* be dreaming any given experience, even paradigmatically normal ones like watching TV in one's living room. That is sufficient to prove the continuity thesis, and it *is* easy to show. Incidentally, it is also sufficient to generate the argument for skepticism about the external world, which was Descartes's main point.

The daydreaming hypothesis about the default network, however, is more ambitious than this weaker version of the continuity thesis, because cognitive functions like planning, memory, and narrative consolidation based on autobiographical information are supposed to function similarly in dreaming and daydreaming. Can one defend a stronger version of the continuity thesis without endorsing the naive version, which is clearly too strong? Descartes famously rejected any stronger view. In particular, he said that the skeptical doubts generated by the dream scenario, which according to many epistemologists are decisive, were based on an exaggeration of what really happens when one dreams. Descartes put forward two theses to defend this claim. The first is that under normal circumstances, when one is awake, one has a strong tendency to *believe* that one is awake. The second is that God would not deceive us into forming systematically false beliefs concerning our wakefulness.[18] Certainly these proposals are not based on *phenomenological* observations, but they do deserve some discussion. We shall focus on the first thesis, and interpret the second as a reliability constraint in terms of reality monitoring.

Reliable memory, or epistemically constrained tracing, is crucial for Descartes's dismissal of the dream doubt. The formation of beliefs concerning whether we are awake rather than asleep is grounded on the capacity to reliably connect current beliefs with past information. The dreamer cannot do this, even if she dreams that she can. Notice that although this claim has some intuitive force, it depends on the degree of *conviction* of the wakeful person, as opposed to the one who dreams. In other words, from "outside" the perspective of the dreamer, one can conclude that she is not reliably connecting her beliefs with past beliefs. But from "inside" the perspective of the dreamer, there is no way to arrive at such a judgment.

Aware of these complications, Sosa (2007, 7) proposes that when one dreams one does *not* believe, in the relevant epistemic sense of the term.

That is, one does not have a doxastic attitude toward contents, equivalent to an assertion of such content, when one dreams. Rather, he suggests, one engages in a form of *imagery*. Sosa claims, therefore, that dream states are not intrinsically identical to wakeful ones because they never give rise to legitimate beliefs. If Sosa is correct, then the continuity thesis fails. Although Sosa is right that there is a substantial difference with respect to reality monitoring, with critical epistemic consequences, phenomenal consciousness in dreams and waking life *is* intrinsically the same, as the continuity thesis holds.[19] How is it then that one *knows* that one is awake? Sosa suggests that the solution is that one *believes* that one is awake, echoing Descartes's proposal. This solution, however, sounds like an assumption rather than an argument. Crucially, monitoring the origin of information is *not* based on beliefs. Belief alone cannot make the required phenomenological distinctions or serve as evidence against the continuity thesis. This generalizes to waking life and dreams. In any case, even if we were wrong about this, it seems clear that access conscious attention is fundamental for understanding how dreams may be related to wakeful experiences, at least at a level that helps to distinguish a waking experience from a dream one.

What are the implications for possible dissociations? If we assume that the unity of consciousness is the same in dreams and waking experiences, then, if the continuity thesis is correct, we should find the same quality of conscious experiences, including conscious attention, in dreams and waking life. The difference will be, presumably, that access consciousness, or access-reality-monitoring attention, is completely absent in dreams. One could interpret this view in terms of the 'field of consciousness' approach. The conscious field would integrate, by means of subsumption, waking and dream experiences in the same way, and operate in tandem with conjunctive attention only in waking life. Notice that this possibility of distinguishing oneiric attention (conscious attention *minus* access attention) and wakeful conscious attention (conscious attention *plus* access attention) entails a Type-C CAD.

One may object, however, that oneiric attention does not really deserve the name 'attention.' Alternatively, one may call phenomenal conscious attention in dreams 'imagery attention' and phenomenal conscious attention in waking life 'epistemic conscious attention.' Since these two types of conscious attention are fully dissociated and one happens without the other, this view seems to entail full dissociation between them. This is one of the most extreme versions of dissociation within the category of 'conscious attention.' While we do not find this view appealing, we want to highlight that it also entails an extreme type of CAD.

4.4.2 The Significance of Oneiric Mnemonic Traces

Oneiric mnemonic traces are not futile, feeble, or utterly isolated cognitive disruptions of an otherwise perfectly working memory system.[20] The previous discussion shows that there are both theoretical reasons and empirical evidence in support of this claim. More important, although mnemonic traces in dreams are not epistemically constrained, phenomenal traces seem to be exactly alike in dreams and waking life, and to comply with the same rules for *phenomenal integration*. If the continuity thesis is correct, oneiric phenomenal traces become important ingredients in the consolidation of narratives for autobiographical memory. Moreover, it also suggests that conscious attention may operate in similar ways in dreams and wakeful states, at least in terms of how one experiences changes in the focus of attention. Clearly, for attention to function properly as an epistemically constrained cognitive capacity, reality monitoring is indispensable.

Dreams, therefore, may be part of the filtering, contextualizing, and assembling of narratives, which is why proponents of the epistemic theories about memory either avoid or criticize the continuity thesis. For instance, Halbwachs, who sought to demonstrate that memory is necessarily a socially mediated process, begins his influential account of collective memory by denouncing the isolated and epistemically deprived character of dreams. Why was this so important for Halbwachs? The answer is that, according to Halbwachs, society is crucial to *epistemically* constrain the contents of memory in a constructive way. Memory, for Halbwachs, necessarily involves wakeful states, but he rejects any *individualistic* theory of memory. Tracing back a memory to its origin is, for Halbwachs, dependent on a narrative that provides the social background to such tracing—a background that is shared and sustained by the individual's community.[21]

Myth and narrative play an important role in this process. Ironically, although myths and collective narratives may originate in dreams and confabulation, Halbwachs distances himself from dreams. By contrast, although Freud agrees that memory is constructive and not neutral with respect to retrieval, he thinks that dreams play an important therapeutic and semi-epistemic role that reveals forms of suppression imposed by society.[22] So in principle, according to Freud, dream memory may play an important cognitive role beyond narrative consolidation; for example, by revealing aspects of repression.

While for Halbwachs society plays a constructive role in the consolidation of personal narratives, for Freud it plays an oppressive role by impeding natural tendencies to form personal narratives. For both, however, narrative is fundamental. Autobiographical narrative is not merely a form of filtering or shaping mnemonic traces that satisfies *epistemic constraints*. An important additional function of retrieving and assessing autobiographical information is to articulate one's own perspective, which cannot be captured by any collection of accurate memories. In any case, oneiric traces play an important role. In the negative account, these traces must be suppressed and systematically distinguished from the collective narrative. In the positive case, they are crucial ingredients to strike yet another balance: the equilibrium between a personal narrative and the narrative that is collectively imposed on us. This balancing role of oneiric traces may be a crucial cognitive aspect of dreaming, one that deserves more careful analysis. Even if dreams prove to be cognitively insignificant, the formation of a self-narrative seems to depend fundamentally on conscious attention, perhaps of the richest kind: *self-aware conscious attention*.

Dream traces are also significant for the transparency method of introspection. If knowing that I have a belief is just my believing such a belief, then if the continuity thesis is correct I should be able to do this in dreams. But it seems implausible that this knowing depends on metareflective knowledge of my situation and of myself as a subject of experiences. Rather, the most plausible view, one for which there is empirical support, is that reality monitoring operates in a less metareflective manner, perhaps largely based on noninferential mechanisms for conscious and unconscious attention.[23] This idea also comports with the phenomenology of self-awareness and experienced effort.

4.5 Summary

Conscious attention is an integral element of human experience, providing us access to perceptual and mental contents that contribute to the experience of conscious awareness. This conscious attention is closely related to the unity of conscious experience that comprises our mental life; it allows for the integration of perceptual inputs as well as past memories, emotion, and imagination. Without conscious attention, one can argue that self-awareness and judgments related to aesthetics and morality could not exist. For this reason, we believe that conscious attention is a recent development in the evolution of cognition, and is thus a special form of attention that is unique to the intersection with phenomenal consciousness and different from the many forms of basic attention that we discussed in chapter 2. We believe it is this conscious attention that affords the richest and most evocative mental contents. These are the mental contents that are necessary for empathic normativity and empathic understanding.[24]

The forms of dissociation that exist between consciousness and attention have direct consequences for how we define conscious attention. From empirical evidence indicating the presence of attentional processes that occur outside of conscious awareness, we must reject any theory stating that consciousness and attention are identical. On the other end of the spectrum, given that there are good reasons to believe that some forms of conscious attention do exist, we must reject the full dissociation view. Now we are left with the tasks of determining which intermediate type of dissociation to accept, and how best to define the overlap between the two that is conscious attention.

In this chapter we have described several forms of conscious attention: those related to phenomenal experiences, including dreams; those related to self-awareness and autobiographical memories; those related to reflexive thoughts (and perhaps to the language of thought); those needed for epistemic seeing (or 'seeing as'); and those related to experiences of effortless attention (such as the experience of flow). Whether or not we can decisively say that conscious attention is a natural kind, however, is still debatable. One way to clarify our understanding of conscious attention is to frame its study in terms of functional purpose, as is often done when describing the evolution of physiological processes in organisms. By examining the possible adaptive role of conscious attention in the evolution of cognitive processes, we should be able to give a better description of conscious attention. This will lead to a better understanding of the relationship between consciousness and attention, and help determine the most probable level of dissociation that exists. It is to this task we turn in the next chapter.

5 Consciousness–Attention Dissociation and the Evolution of Conscious Attention

Herbert Spencer famously wrote that "If the doctrine of Evolution is true, the inevitable implication is that Mind can be understood only by observing how Mind is evolved" (Spencer 1855, 291). Given the importance of the theory of evolution in scientific explanations of biology and psychology, this statement appears incontrovertible. The last few decades of research in psychology and neuroscience, however, show that it is an ambiguous assertion. The conscious mind has proved elusive of evolutionary explanation, so elusive that many theorists argue that phenomenal consciousness *cannot* be adaptive because it has no specific function (Chalmers 1996). Although the consensus is that Spencer's claim must hold for most cognitive processes, phenomenal conscious experience seems to escape functional explanations, and this has consequences that are not properly understood.

We now focus on one of these consequences: the tenuous account of the relationship between consciousness and attention. In this chapter, we argue that the scientific findings about attention and basic considerations about the evolution of different types of attention demonstrate that consciousness and attention must be dissociated regardless of which definition of these terms one uses. To the best of our knowledge, no extant view on the relationship between consciousness and attention has this advantage. Because of this characteristic, we can now present a principled and neutral way to settle disputes concerning this relationship, without falling into debates about the meaning of consciousness or attention. A decisive conclusion of this approach is that consciousness cannot be identical to attention.

While most cognitive psychologists agree that selective attention must have been an early adaptation in the evolution of cognitive systems (Ward 2013), few would consider *consciousness* to be an early adaptation. Rather, it may be an evolutionary development that appeared later in the progression of cognitive systems. In this sense, consciousness may help serve the purpose of controlling integrative aspects of selective attention (Rensink in press), although this possibility has yet to be fully clarified. Regardless, a longstanding claim in the literature is that evolutionary considerations are useless for an understanding of consciousness because conscious awareness seems to lack any specific function, especially since integrative processes in attention can occur without conscious awareness (Mudrik, Faivre, and Koch, 2014, Talsma et al. 2010, Zmigrod and Hommel 2011). Some authors even claim that consciousness might be a spandrel—that is, a by-product of the increasing complexity in brain structures that does not serve any evolutionarily adaptive purpose (Carruthers 2000, Dennett 2005, Gould and Lewontin 1979, Polger and Flanagan 2002, Rosenthal 2008). This creates a fundamental problem in our current understanding of consciousness because it seems to prevent a functional analysis of *conscious attention*, that is, the reportable type of attention that is part of conscious awareness.[1] This specific problem has not been given a sustained treatment in the literature, and it has decisive consequences for the relationship between consciousness and attention.

One of the most important implications of this evolutionary question is that the consensus found in the literature on the nonadaptiveness of phenomenal consciousness favors a strong dissociation between consciousness and attention. That is, their relationship is not one that coevolved such that the two are linked functionally. Surprisingly little published research has focused on the evolution of attention, especially in terms of how it relates to the evolution of consciousness (but see Cosmides and Tooby 2013, Tooby and Cosmides 1995, Ward 2013, Wright and Ward 2008, 235–41). Speculations about the evolution of consciousness, on the other hand, have received more treatment in the literature (e.g., Edelman, Baars, and Seth 2005, Feinberg and Mallatt 2013, Nichols and Grantham 2000, Polger and Flanagan 2002, Seth and Baars 2005).

The role of evolution, therefore, becomes a central consideration against theories that fully identify consciousness with attention. As the amount of experimental evidence against the identity view continues to grow (e.g., Dehaene and Naccache 2001, Lamme 2006, van Boxtel, Tsuchiya, and Koch 2010), the evolutionary approach provides a principled theoretic way to explain why scientists are finding such dissociations. Thus, instead of arguing for dissociation based on inductive arguments from those specific bodies of evidence, we argue instead for a more decisive and principled rationale for dissociation, based on the importance of evolution in scientific explanations. The unique advantage of this theoretical approach is that it is interpretation independent. Regardless of how 'consciousness' or 'attention' are defined within the theories currently available, there will be a substantial dissociation between them.[2] This argument avoids the problems that derive from the multiple definitions of the terms, obviates semantic disputes, and puts the current evidence under a much clearer theoretical light.

Our approach to the debate on the relationship between consciousness and attention is decisive in at least two respects. If it is true that consciousness has no cognitive function, then the type of dissociation between consciousness and attention must be either the most severe (full dissociation, or independence) or very severe with very few cases of conscious attention (Type-C CAD). Alternatively, if consciousness has a specific evolutionary function, and theorists seek to provide an identity view or even a mild dissociation view (Type-A CAD), then they will face the challenge of specifying how consciousness as a cognitive function coevolved with attention, which can be defined in terms of *specific* types of cognitive functions, such as voluntary, object-based, feature-based, or spatial attention. Given our current theoretical and empirical understanding, we suspect that this latter possibility will not materialize. We argue that a major complication of identifying consciousness with attention is that empirical findings suggest that different types of attention must have evolved at different times, making an identity argument difficult to defend.

5.1 Attention as an Early Adaptation

In their influential paper on the evolution of cognitive functions, John Tooby and Leda Cosmides (1995, 1195) argue that attention must be one of the earliest adaptations in the evolution of the human mind. Although they talk about perception more generally in the same passage, it is clear that they also include basic low-level attentional processes, such as those involved in navigation and feature detection. Indeed, one can hardly think of a more basic cognitive function than selective attention for motor control and navigation. Many animals navigate by exploiting features of the environment—locating the position and angle of the sun to determine orientation, for example, or performing computations on how locations of external features relate to their egocentric frame of reference. Beyond such observations, there has been minimal examination of the evolution of attention; only brief discussions are found in the current literature (e.g., Cosmides and Tooby 2013, Ward 2013, Wright and Ward 2008, 235–41).[3]

A recent review on evolutionary psychology by Cosmides and Tooby (2013), however, identifies several relevant themes regarding visual attention, spatial navigation, and social interactions. From this perspective, the related attentional mechanisms—such as the spatial attention and memory used in hunting and foraging for food—are thought to have evolved because they increased chances for survival by monitoring critical aspects of the environment. The authors cite several studies indicating that even humans have a better memory for food locations. Additionally, some forms of attention seem to be adapted for social behaviors that emerged in communal species like humans, such as mate selection and kin detection (the nuanced identification of in-group members and faster recognition of anger in out-group members), for 'welfare trade-offs' (balancing self-interested and altruistic behaviors), and for the motivation that drives these social enrichments. These adaptations evolved later than the more basic ones like feature-based attention or spatial attention related to navigation.

Similarly, the effects of emotions on an organism can take several forms, which help determine behavior. Joseph LeDoux (2012) describes the fundamental forms of emotion, upon which more complex human emotions may be based, as 'global organismic states' that prepare an organism for making an appropriate response to environmental factors. Fear, for instance, is a fundamental motivation that helps draw attention to crucial aspects of the environment that affect survival. The more socially mediated emotions emerged later, along with social interactions, and these 'feelings' have a phenomenal character independent from the more basic 'global organismic states,' which are related to survival circuits and are closer to what others would describe as access consciousness (LeDoux 2012). This supports the idea that there is a dissociation in the evolution of these different mechanisms that regulate attention to, and awareness of, critical environmental factors.

As discussed in chapter 2, spatial and feature-based attention are fundamental for performing tasks critical for survival. Parsing the environment into objects and spatial coordinates, or into predator/prey and conspecifics, require these forms of attention. These basic forms of attention can

CAD and the Evolution of Conscious Attention

interact with other cognitive capacities such as short-term and long-term memory to produce richer representations. Animals with robust nervous systems can use episodic and semantic memory to recognize targets in their environment and increase the repertoire of objects and features that can be attended at any specific time. These capacities, in turn, become the basis for optimal decision making, task switching, and action planning. The latter are behaviors that animals display in their environment which require a fair amount of cognitive integration. It is likely that skills related to attention were critical for survival as competition for resources intensified during the Cambrian era. Incidentally, some authors, particularly in the biological sciences, argue that the increase in information complexity that accompanied integrative processes of attention and memory is an indicator of the emergence of consciousness (e.g., Nichols and Grantham 2000). The problem with this proposal, however, is that the functions of feature integration, as well as the higher complexity that relates to conscious experience, can be fully understood and explained by attention without phenomenal awareness, a point we develop later in this chapter.

Competition and the struggle for survival—the drives underlying evolution—require that species develop some form of interfacing agent-based attention, with attentional mechanisms selected because of their automaticity as immediate responses to the environment. Reflex-like reactions to features of the environment that immediately trigger or grab attention must therefore be interfaced with processes of associative learning that initiate with sustained voluntary attention and eventually produce more complex behaviors that can become automatic (as described by our discussion of 'effortless attention' in chapter 2). Success in search tasks and behavioral routines, like those associated with foraging or hunting, becomes a staple of the fitness of a species. The degree of cognitive integration required for this interaction between the different forms of attention is significant, and marks an important transition in the way in which information is stored—a transition in informational complexity, as defined by Smith and Szathmáry (1995). Furthermore, the integration required for search behavior also demands short-term and long-term memory, as well as mappings between frames of reference that constitute the egocentric perspective necessary for creatures with a central nervous system (see Merker 2005).

Thus many forms of attention must be adaptive and must have evolved very early, as organisms diversified and developed central nervous systems in order to cope with ever more complex and demanding interactions with their ecosystems. From basic discrimination to highly integrated search behavior, a repertoire of attentive skills must be in place in order for a species to adapt and survive. Crucially, these skills do not constitute a ladder from lower forms of cognitive systems to higher forms, nor are higher forms of cognitive complexity causally necessitated by lower forms. Rather, the transition to complex forms of attention is best captured in terms of intentionality and of feature binding into higher-level object representations. Creatures with the capacity to integrate representations with content about the environment, its objects, and its spatiotemporal structure are capable of many other functions that depend on attention and representation, such as learning and remembering information. Communication and language capacities are similarly related to this form of high-level symbolic representation.

One can illustrate the previous point by examining the evolution of the sensory modalities. Animals with a nervous system capable of registering light waves and encoding that information in a representational fashion can cope with many tasks that depend on attending to visible features of the environment. In this sense, intentionality—the 'aboutness' of cognitive states that is directed toward something—must be present at some level for goal-directed behavior. One can imagine a creature that is capable of at least selective attention for object and feature recognition within a sensory modality equivalent to human vision. Such a visual sensorial apparatus would not be inferior to the more complex crossmodal sensorial registration that characterizes mammals. Crossmodal integration, however, allows animals to significantly enhance the confirmation of information across different encoding systems, like light or sound waves, and interactions between different modalities can facilitate perception, as for example by enhancing motion perception through the localization of sounds associated with an object (Zmigrod, Spapé, and Hommel 2009).

Perceptual questions that the brain needs to answer within a single modality—is this the same object that was moving a second ago?—can be formulated within a broader network of possibilities, such as the simultaneous instantiation of properties detected crossmodally from a single distal stimulus. Not only spatial attention, but temporally sustained attention becomes more flexible and dynamic. It is known that each modality has a unique simultaneity window and that these windows differ considerably

(Pöppel 1988). These varying windows serve to compensate for the speeds of the energies that they are designed to register, and are part of a cross-modal window of simultaneity required for motor control tasks (see Montemayor 2013, 95–112). The evolution of different modalities, therefore, led to a flexible, dynamic, highly integrative, and evidence-driven centralized system for information processing.

Such arguments indicate that one cannot reasonably deny that attention, in its many forms, is an adaptation. The key point, however, is that these forms of attention are defined *functionally,* as we described in chapter 2, while there are powerful arguments against functional accounts of phenomenal consciousness. The cognitive functions associated with attention are fundamental adaptations in the course of the evolution of the central nervous system, and yet none of these functions seem to require conscious awareness, a point on which many researchers agree (e.g., Dehaene and Naccache 2001, Lamme 2006, van Boxtel, Tsuchiya, and Koch 2010).

Although selective attention is certainly an early adaptation in the evolution of the human brain, the claim that *conscious awareness* is an adaptation is considerably more controversial. The functional-computational approach assumed by evolutionary theory and cognitive psychology has been criticized by many theorists as useless to account for conscious awareness (e.g., Block 1995, Chalmers 1996). Even if one defines consciousness functionally—as in the global workspace account (Baars 1988, 1998, Dehaene and Naccache 2001), or the integrative account (Tononi 2008, 2012), or in an account that emphasizes learning (Meuwese et al. 2013)—one can still show that attention is an early adaptation and that the alleged functions of consciousness must be a much later addition to the repertoire of cognitive capacities. This *early* versus *late* adaptation implies dissociation. Effectively, such arguments demonstrate dissociation and preclude an identity account of the relationship between consciousness and attention.[4]

Furthermore, there seems to be consensus that consciousness appeared recently in the evolution of cognitive capacities. While there are several accounts of the evolution of consciousness, ranging from complete skepticism to proposals that consciousness may be a spandrel, *none of these accounts claim that consciousness is an adaptation.* Cognitive scientists are therefore confronted with the puzzling challenge of reconciling conflicting claims concerning consciousness and attention into a theoretically unified account of the evolution of conscious attention.

In what follows, we present a theory for the evolution of conscious attention that entails a substantial, but not full, dissociation between consciousness and attention, which according to our classification would be at least a Type-A or Type-B CAD. This conclusion is in sharp contrast to recent views that try to identify consciousness and attention. The main conclusion of this chapter is that consciousness and attention must be largely dissociated for evolutionary reasons. Specifically, the degree of dissociation is severe because while most forms of attention must have evolved early as adaptations, no form of conscious awareness was required as an early adaptation; this suggests that the overlap between consciousness and attention is an equally recent phenomenon and must be the exception, rather than the rule, with respect to the interaction between consciousness and attention.

5.2 Attention, Functionalism, and Evolution

The vast research on the different forms of visual attention confirms that attention should be defined functionally. In fact, definitions concerning the types of attention aim at *capturing* different functions (see Rensink, in press). Selective attention in its most basic form has the broad function of filtering information and distributing cognitive resources for optimal modulation of information processing. The idea that attention performs a kind of filtering function is central to our understanding of it, and the first scientific theories of attention defined it in terms of monitoring and filtering, as we discussed in section 2.1.2.

There are different types of visual information, however, that can be selected by low-level attentional mechanisms that focus on feature or spatial information. As we described in section 2.1, feature-based attention selects various types of perceptual features in a visual scene, such as color, segment orientation, or motion. Spatial attention has the function of quickly parsing the surrounding environment into regions for information registration. Object-based attention is a midlevel form of attention that specifically targets units of discrete objects, and the mechanisms for detecting, tracking, and counting objects have been found in many species. Using Spelke's (1994) terminology, these forms of attention are constitutive of an ancient and fundamental 'core knowledge' that structures many types of cognition and information processing. This core knowledge is observable in young infants and facilitates an understanding of physics, psychology, number, and geometry. While all forms of attention are defined in terms of functions related to information processing, some forms of attention are more basic than others. In this section, we argue that the current theoretical and empirical understanding of attention allows us to hypothesize that different forms of attention evolved at different times. This thesis has critical consequences for the evolution of conscious attention and for views that identify consciousness with attention.

5.2.1 Selective Attention as the Earliest Form of Attention

Once the functions of attention are identified, the neural correlates of those functions can be studied across species. For example, selective attention and object-based attention, capacities that include basic discrimination functions, involve substantial portions of the cortex in humans. As Ward explains: "Many brain imaging studies show that even apparently simple tasks, like shifting attention from one object to another, activate a large network of cortical areas" (Ward 2013, 54).[5] Shifting focus between objects is a fundamental function of attention, indispensable for the selective interpretation of visual information without which basic navigational or visual search tasks would be impossible. Neural studies on humans and nonhuman primates also identify a hierarchical organization of visual areas that increase in processing complexity. This hierarchical organization mirrors the increasingly complex functions of vision that correspond to the different forms of visual attention required to select and integrate information within the hierarchy, as discussed in section 2.2.

From an evolutionary perspective, the ability to attend to multiple features of the environment is so ubiquitous that it must be physically instantiated by different neural correlates across species, including insects. Indeed, forms of selective attention to environmental features have been found in many species, including some with minuscule brains such as fruit flies or dragonflies (Wiederman and O'Carroll 2013). While it is clear that insects do not enjoy the attention capacities of humans and animals with larger brains, this lower-level common ability is extremely relevant, because fruit flies diverged 500 million years ago from the human branch in the evolution of different types of nervous systems (Dawkins 2004, Ward 2013). This form of selective attention must be causally driven. It depends on automatic processes that detect features according to principles of salience, and is selective in the sense that such features trigger some sort of relevant goal-related behavior that results from adaptive influences, or can become available for selection by the filtering mechanisms of attention.

The multiple correlates of the functions of basic selective and object-based attention open up intriguing possibilities. For example, the cortex is a large area of the brain, costly in terms of energy expenditure. So basic forms of selective attention must be physically instantiated in much smaller and less costly brain areas (see Ward 2013), which suggests that forms of selective attention likely evolved at different times in different species with very diverse neural anatomies. In fact, it could be argued that the presence of such a variety of neural systems that support attention-related processes suggests that selective attention has evolved repeatedly and independently in different species, as is generally believed to be the case for the evolution of various brain systems (see Striedter 2006).

We must emphasize that we are not claiming that these forms of cross-species attention are identical in all respects. It seems obvious that one reason why attention activates vast areas of the human cortex is because of the rich and highly integrated contents that human attention is capable of processing. We do not believe, for example, that any and all forms of signal detection qualify as selective attention. Rather, we are noting the empirically confirmed fact that many species with smaller nervous systems are capable of shifting attention from one object to another, as insect navigational capacities demonstrate (see Gallistel 1990a for discussion about navigational functions that depend on such capacities, like the solar ephemeris function). These cognitive functions are verifiable empirically and are possible only because of the integration of the information being represented and attended to, rather than due to the mere causal interaction between internal biology and external factors. This kind of attention, manifested in simple shifts, must be among the most ancient of cognitive functions. We remain neutral, however, with respect to the demarcating question of which species do or do not have attention capacities similar to human attention.[6]

The likelihood that selective attention evolved repeatedly and independently across many species suggests that it is a crucial and very basic adaptation (Ward 2013). Following Smith and Szathmáry's (1995) proposal concerning the evolutionary importance of transitions for the way information is stored and used, the appearance of cognitive systems capable of discriminating and selectively attending to different features certainly must have marked a transition into more integrated and complex behaviors. At a minimum, selective attention reveals purposeful behavior and a minimal form of intentionality, susceptible of a teleologically structured semantics according to which mental contents are specified by the conditions required to succeed in pursuing goals. As more complex forms of object recognition became available, for example by focusing attention on conceptual content retrieved from long-term memory, better problem-solving and planning abilities became available. Yet the resources of selective attention continued to rely on those same early adaptations that animals need for such basic survival activities as navigating through their environment.

Parenthetically, it is productive to contrast the pervasiveness of attention as an early adaptation with the scarcity of what some theorists have suggested to be the basic functions that underlie the spandrel of conscious awareness: the capacities for mind reading and language (Carruthers 2000, Dennett 1969, 2005). Language may itself be a spandrel and a *uniquely human* capacity, according to some theorists (Fitch, Hauser, and Chomsky 2005). Mind reading, or at least having a rudimentary theory of mind, seems to be more widespread, but robust forms of mind reading also seem to be uniquely human, such as those required to pass the false-belief task or to have socially specified forms of attention to detect the intentions of conspecifics (Tomasello 1999).[7] Notice that these views do not consider phenomenal consciousness to be an illusion, a mystery, or a primitive feature of the universe.

Even if cognitive functions could explain the emergence of consciousness, it would still be dissociated from attention, since the lower-level forms of attention also have evolved in organisms that do not seem to have any conscious awareness (see Griffin and Speck 2004). This is not to deny that there may be phenomenal consciousness in animals. Some animals with central nervous systems may experience pain or color in ways that resemble human experience; yet the functions of attention would precede or at least be evolutionarily independent from these neural processes. This conclusion does not depend on how consciousness itself is defined. If one defines consciousness functionally, one could think of it in terms of 'access consciousness,' using Block's (1995) distinction. Access consciousness would be dissociated from early forms of attention in the same way, and issues surrounding non–functionally defined phenomenal consciousness would just make the dissociation even more severe. We expand on these issues in section 5.3. In the remainder of this section, utilizing our detailed review in chapter 2, we summarize how the empirically accepted forms of attention must have evolved at different times and describe how they relate to mechanisms for crossmodal integration.

Feature-based attention, as discussed in section 2.1.1, can be characterized as a primitive information-selection mechanism that interacts with low-level perceptual processes. There is evidence that this type of basic-level attention appears among both invertebrates and vertebrates, including dragonflies, octopi, fish, frogs, rats, and primates (Maunsell and Treue 2006, Ward 2013, Wiederman and O'Carroll 2013). Generally, this is a bottom-up form of attention that operates automatically within complex and feature-rich environments in ways appropriate to its role in the evolutionary development of these systems (Rensink, in press). These systems are hierarchically organized, such that different levels in the visual cortex are responsible for certain features, and a biasing of the competition for feature selection can occur on several of these levels through feedback loops that integrate the influence of higher-level attentional processes. These information processing systems are structured by the modular organization of specialized brain regions responsible for registering specific types of visual information, such as color, motion, or segment orientation, with the processing of basic features occurring in 'lower' areas of the brain (see section 2.2). This organization is related to the 'modularity of mind' hypothesis (Fodor 1983), which suggests that different perceptual modules developed for processing different features; although recent studies indicate that they are not distinct brain regions and may be better described as organized into distributed networks (see Finlay and Brodsky 2006).

Feature-based attention interacts with low-level sensory systems to select information in a typically automatic way, but this selection process can be biased by top-down signals determined by task demands. As described in section 2.2, the specialized regions in the visual cortex that process specific types of visual information are accessed by attentional mechanisms in different ways by separate neural structures. There are also distinct processing streams for visual information that affect behavior and influence what enters conscious awareness—the ventral and dorsal streams. One could even argue that the dorsal 'vision for action' pathway is evolutionarily older than the ventral 'vision for perception' pathway (Milner and Goodale 1995). The results of these independent feature-specific processing mechanisms, which then may be organized into feature maps, can be

CAD and the Evolution of Conscious Attention 189

linked together based on spatiotemporal information to form more detailed representations (see Treisman and Gelade 1980, Treisman and Zhang 2006). Feature-based attention can be considered a fundamental form of attention from which richer representations are built.

Another basic form of attention, spatial attention, filters information based on spatial coordinates and must have evolved early in order to aid navigation and other goal-directed motor behaviors. Spatial attention can operate on discrete objects or empty space, or spread globally to quickly determine the gist of the information present across the visual field. Indeed, a very important ability of distributed spatial attention is to quickly capture a statistical summary representation of the information that lies beyond the focus of attention (Alvarez and Oliva 2008). This observed ability, which allows visual information to be pooled into summary representations and thus helps to overcome the visual processing limits imposed by attention-demanding tasks, points to a more fundamental purpose of spatial attention.

The neural systems that support spatial attention are located in areas related to motor activity (Beauchamp et al. 2001), and thus can be considered more primitive. Although classifications of neural systems supporting different aspects of spatial attention are still under study, we do have some sense of how different processes are localized in different brain regions and that understanding distinguishes these forms of attention based on their functions. For example, more complex spatial abilities such as a task-specific rehearsal of spatial information are present in vertebrates and appear to be concentrated in the right brain hemisphere, with the frontal and parietal structures being particularly active during spatial attention tasks, and these structures also have some overlap with working memory networks (Awh and Jonides 2001).

More specialized shifts of attention, such as covert attention, seem to require neurons that are involved in executive control and are found only in higher primates such as apes and humans (Wright and Ward 2008, 240). Covert attention is the capacity for a willful shift of attention outside the center of one's gaze without changing the direction of gaze. That is, the focus of attention can be oriented independently of physical manipulation of the gaze—in contrast to overt attention, which refers to what is at the center of attention and usually on the fovea (center) of the eye, as discussed in section 2.1.1. The shifting of covert attention occurs higher in the brain hierarchy, in the frontal eye field of the frontal cortex (Thompson, Biscoe, and Sato 2005). Such specialized attentional shifting may also be involved in the ability to attend to certain thoughts from memory or to other mental states that are not immediately linked to sensory information, a function that indicates a more advanced use of attention.

Attention can also be drawn to things in the world that display object-like properties such as cohesion, symmetry, or common fate (see section 2.1.1). This object-based attention requires a two-stage process that begins with the bottom-up *individuation* of objects, upon which a selective attention operates such that object features are bound and maintained into a persistent representation. This binding can then lead to the *identification* of the object. Together, individuation and identification contribute to the experience of attending to specific items in the world.

The first stage of object-based attention, individuation, is a basic data-driven process in early vision that selects discrete objects for further processing and is thought to occur in parallel (Pylyshyn 1989, 1999, 2001). This process of individuation allows visual references or 'pointers' to be assigned to individual objects in a visual scene, which can be maintained over temporal and spatial changes.[8] The second stage in object-based attention is identification. This stage requires the binding of object features from feature maps into sustained mental representations that enable object identification and recognition (Treisman and Gelade 1980). A selective focused attention plays a crucial role in forming persisting 'object file' representations by allowing features from a visual scene to build a coherent representation incrementally in visual working memory (Kahneman, Treisman, and Gibbs 1992, Treisman 1998, 2006). Given that these are capacities for basic object perception and that animals tend to interact with discrete objects like food, mates, or basic tools, similar forms of visual mechanisms for individuation and identification must be present across species.

The object-based attention model argues that visual cognition tends to operate on whole objects in a visual scene, as opposed to operating on individual features—a claim supported by the many studies we described in section 2.1.1. Such object-based representations are formed in visual short-term memory, which has a context-dependent capacity limit regarding the total number of objects and the amount of featural information that can be represented (Alvarez and Cavanagh 2004). This more complex form of attention is related to voluntary endogenous attention and requires selective focus to bind and maintain the features of perceived objects over time. In light of the more centralized voluntary aspects of this form of attention, its evolution probably does not correspond with the earlier appearance of the most basic and adaptive forms of attention, like feature or spatial attention.

In a sense, the visual indexing, feature integration, and object file theories we have described provide an account of the way detailed object representations are formed through the spatiotemporal binding of featural information. These theories describe the integration of independently processed visual features to produce representations that refer to external objects. It is these object representations that are most likely the contents of conscious experience, even though not all object files reach conscious awareness (Mitroff, Scholl, and Wynn 2005). Note, however, that this is a claim about *content*, rather than about the basic functions of early visual attention. A major problem with the characterization of the function of consciousness is that these contents may be processed without any associated *experience* of the contents—that is, there may be no associated feeling, phenomenal property, or experience of 'what it is like' to be in a state representing those contents.

As described in section 2.2, brain areas supporting object-based attention are generally considered to be more highly evolved. In addition to the visual cortex sending inputs and receiving feedback signals (Cohen and Tong 2013), the superior parietal lobe above the visual cortex and the right lateral fusiform cortex seem to be involved in object-based attention, including the shifting of attention between objects (Yantis and Serences 2003). Object tracking and motion detection are supported by the middle temporal lobe (Newsome and Paré 1988), and related actions engage suppressive interactions from the extrastriate areas V2 and V4 (Bundesen, Habekost, and Kyllingsbaek 2005, Desimone and Duncan 1995, Luck et al. 1997, Mangun 1995, Posner 1992). These structures have been identified in other species, including pigeons and monkeys, but only a few studies have investigated object-based attention in animals (see Roelfsema, Lamme, and Spekreijse 1998, Ushitani, Imura, and Tomonaga 2010).

To reiterate, our focus here is on attention as a primitive cognitive function, and on why this function is one of the earliest adaptations of the nervous system, as the vast evidence on spatial and feature-based attention for basic navigational skills demonstrates. So we are not claiming that the contents that humans are capable of attending to are also part of the early adaptations that constitute the functions of attention in early vision. As we have argued, contents, especially conceptual ones, are likely to be related to conscious attention and may not be necessary for basic discrimination tasks such as feature-based attention.

Those who defend the view that attention is identical with consciousness must either say that any animal capable of navigating and selecting features from the environment is conscious, or claim that these basic forms of information processing do not deserve the name 'attention.' Because of the evolutionary considerations we are using as theoretical background, as well as the broad consensus that these basic forms of attention are empirically confirmed, we find both options highly problematic. These early forms of attention are fundamental and adaptive, giving rise to crossmodal attention as a later adaptation.

5.2.2 Crossmodal Attention

As has been emphasized in the vast literature on visual attention, the recognition of objects and their features within the visual modality requires mappings based on spatiotemporal coordinates. The feature integration theory for visual object recognition is based on this idea (Treisman and Gelade 1980). Although there is controversy concerning this proposal (Pylyshyn 2007), it is uncontroversial that human visual attention is capable of integrating a vast number and variety of features that are spatiotemporally organized into coherent and useful representations. Crossmodal attention in nonhuman cognition need not be as rich in feature integration as it is in humans, but it certainly requires mappings between different geometric structures such as those involved in vision and audition. Proprioception and crossmodal attention, therefore, seem to require a centralized nervous system and a centralized way of integrating and interpreting different types of information, which may be the basis of voluntary attention.

In the same vein, Hommel's event file theory proposes an enhanced version of object files that incorporates both crossmodal features and motor commands (Hommel 2004, 2005, 2007b, Zmigrod, Spapé, and Hommel 2009). This theory provides a richer notion of object representations, in the sense that these crossmodal sensory representations also integrate action-planning information. There is also support for early interactions across modalities in a task-irrelevant perceptual learning of crossmodal cuing effects (Batson et al. 2011). That is, when an irrelevant sound is paired with a specific stimulus in a manner not obvious to the test subject, subsequent post-training performance can nevertheless be affected in a way that indicates a crossmodal association was implicitly learned. In order for these associations to occur, information from different modalities must be integrated in some way, and this necessitates some sort of mechanism allowing the integration to occur.

Crossmodal attention integrates a variety of features and requires mappings among the different sensory modalities. One of these mappings is based on attended objects and features across modalities. Another mapping concerns egocentric and allocentric frames of reference that enable an organism to navigate successfully within an environment (Gallistel 1990a, 1990b). With respect to time perception, these mappings create the need for a crossmodal window of simultaneity, because each modality has been found to have a different simultaneity window (see Montemayor 2013, Pöppel 1988). The basic idea here is that integration of crossmodal information and robust integration of memories, both episodic and semantic, create a platform for the egocentric perspective that seems to characterize phenomenal consciousness.

Like the previous cases of attention within a single modality, crossmodal attention for features, objects, and regions of space is likely to have been adaptive, as species with brain areas capable of mapping information from one region to another created more robust, complex, and stable representations of their environments. Crossmodal integration seems to be the basis of many processes that increase the probability of success at tasks crucial for survival, such as foraging and hunting. While this seems obvious, it is significant that these mappings require sophisticated representations of the environment, or at least more robust forms of memory for space and feature binding than those required by more basic forms of attention. Such processes, however, need not entail that the representations or cognitive capacities involve conceptual content. Nevertheless, these crossmodal mappings and integration processes indicate the presence of a more highly evolved cognitive system. Even if conscious attention serves this integrative purpose, it would still be a system that evolved at a later stage after the early forms of attention appeared. As we discussed in section 3.2, cognitive integration does not necessarily produce conscious awareness, although integrative theories proposed by Tononi, among others, might suggest it can.

Because of the theoretical and empirical reasons outlined in this section (and detailed in chapter 2), our current understanding of attention allows us to hypothesize that different forms of attention evolved at different times. An account of the evolution of semantically based representations in the context of more evolved forms of attention and the emergence of voluntary attention reinforces this view.

5.2.3 Conceptual Content and Attention

The emergence of semantics and conceptual content is a notoriously controversial issue in philosophy and cognitive science. Some philosophers think that only humans have the capacity for conceptualizing and forming beliefs based on concepts, and to endorse beliefs based on such concepts.[9] Conceptual content seems to be deeply associated with the language capacity because of its compositional and generative characteristics (Fodor 1998). It is also associated with more complex activities of imagination and metaphor (Lakoff and Johnson 1980). All theories of conceptual content relate it to the language capacity or to the capacity to form beliefs that support inferential reasoning; therefore, given the assumption that language and inferential reasoning are recent phenomena in the evolution of our species, conceptual attention seems to be a recent phenomenon.

Are conceptual contents uniquely human, or do other species have them? Without endorsing any specific view on the scope of the conceptual contents of attention, we would like to illustrate the type of cognitive enrichment that highly integrated conceptual contents enable and also to emphasize their relatively recent evolution. For example, contents are conceptualized in Gestalt phenomena and ambiguous images in general. In cases like the Necker cube, the image is conceptualized as a three-dimensional cube and can alternate in appearance, from having an upward orientation to a downward one, generating an attention pattern with a specific temporal rate. In other cases, however, the attention pattern is between two very *different* conceptual contents, as in the drawing that can appear to be a rabbit or a duck, depending on how the image is interpreted (see figure 1.2). This kind of conceptual attention shifting must have appeared at later stages in the evolution of attentional mechanisms. Although they depend on automatic processes, like the back-and-forth switching process characteristic of the perception of ambiguous images, these contents of attention are clearly much more conceptualized than the objects of basic forms of selective attention, which constitute structural features of the visual scene or of other perceptual modalities.

Similarly, enrichments of skill that build on reflex-like capacities must also be based on attention to the conceptual content of perception. This is achieved by gathering perceptual information and comparing the unified representation of these groups of features to concepts stored in long-term memory. Object-based attention provides a feature integration mechanism that enables the recognition of objects, people, and places in the external world. In a very simplistic account, being able to recognize things in the world relies on this binding of features—both basic features, such as shape and color, and conceptual elements, such as seeing something *as* a duck or a rabbit. More realistically, however, these basic processes are supplemented by the rich information stored in memory. For example, the detection of only a few features is often sufficient for the brain to predict the identity of the object based upon the stored content in long-term memory. Conceptual cognition relies on basic perceptual processes and has evolved to make possible the interaction between these basic processes and the stored content in long-term memory, increasing the efficiency of perception and cognition. This advancement necessitates a complex integration of not only perceptual but also conceptual information, a process that integrates not only currently attended features but potentially relevant ones. The presence of such concepts in animals, which must exist on a very basic level, can serve the purpose of providing representational templates or schemas for recognizing features and objects that are relevant for survival (Cosmides and Tooby 2013).

5.2.4 Voluntary Attention

Sustained attention is deeply associated with the passage of time and the sense of self. For this reason, it seems that voluntary (or willfully sustained) endogenous attention must be at the basis of self-awareness. Here an interesting issue emerges. There must be two kinds of voluntary attention, with and without self-awareness, because many forms of sustained attention, like keeping track of moving objects or searching for a particular feature among distractors, can be performed without high levels of self-awareness. But clearly, self-awareness is associated with sustained attention, as when one meditates or recites a phrase to oneself to consolidate memories. A simpler form of sustained attention must have emerged early on, and is closely related to the selective attention processes described earlier that support object tracking or focusing on details. The form of voluntary attention associated with the sense of self and the flow of time, however, must be more recent, because they require self-reflection and metacognition, which necessitate more complex neural connections.

Voluntary attention is contrasted to the more automatic and primitive *involuntary* forms of attentional control. The distinction between involuntary and voluntary attention is often considered to be the distinction between bottom-up and top-down attention, as we noted in section 2.1.2. That is, attention can be stimulus-driven and automatically guided toward important external events that involuntarily attract the focus of attention, or it can be voluntarily guided through willful selection. Forms of involuntary attention also can bias neural activity toward selecting information into higher processes and can affect behavior without reaching conscious awareness—a finding that supports a dissociation between attentive processes and consciousness.

From an evolutionary perspective, bottom-up involuntary attentional processes can be seen as more primitive, earlier adaptations that aid an organism in basic evolutionarily relevant survival tactics such as procuring food, avoiding prey, protecting kin, or finding appropriate mates (Cosmides and Tooby 2013). These early perceptual selection processes are tied to feature and spatial forms of attention in which information is processed according to the specialized functional modules in the brain. Those brain regions emerged early in evolution and are present in invertebrates, animals that usually would not be considered to have conscious awareness.[10]

On the other hand, voluntary attention can be described as being endogenous and more deliberate, and it can bias the competing neural activity in lower-level processing to support the goals of the current task. Voluntary attention is likely to have evolved later as organisms were required to adapt and respond to more complex and integrated representations—the challenges involved in tool usage, various forms of learning, and social interactions such as communicating and maintaining social order within groups. These higher-level cognitive processes require a voluntary sustained form of attention, as well as interactions with other cognitive processes such as working memory and long-term memory. Animals exhibiting such higher-level behaviors include blue jays, crows, parrots, macaque monkeys, and chimpanzees, all of which show signs of tool manipulation or rudimentary language abilities (Griffin and Speck 2004). Having self-awareness, however, seems to be a yet more complex function of voluntary attention than mere sustained attention: it involves reflecting not only on the contents of what is being attended but on one's own conscious perspective as well. That is, one is not only aware of the contents, but also aware that one is thinking about them.[11]

Various brain regions considered to be more evolutionarily advanced have been identified as supporting voluntary and top-down attention, as we noted in section 2.2. To briefly reiterate, these include areas like the prefrontal cortex, also associated with executive function (Baddeley and Della Sala 1996, Chica, Bartolomeo, and Lupiáñez 2013, Petersen and Posner 2012). Parietal and frontal regions in the cortex participate in attentional control by initiating and maintaining attentive states related to specific goals (Yantis and Serences 2003), but are constrained by low-level computations in the visual system (Yantis 2000)—demonstrating how top-down and bottom-up processes interact. Similarly, recent studies suggest that feedback and recurrent processes explain how top-down modulation occurs, which may be closely tied to conscious attention (Lamme 2006, Pollen 2003, Theeuwes 2010). The anatomy of the brain and its evolution, therefore, strongly suggest that different forms of attention interfaced and became interlocked at different stages in the development of our species. This organization also indicates that more advanced forms of attention, related to conscious attention, corresponded to the growing complexity in the cortex and the presence of recurrent pathways to the "older" brain regions.

5.2.5 Effortless Attention and Flow

Effortless attention, like bottom-up attentional processes, is considered an involuntary, sensory form of attention that does not always reach conscious awareness. These effortless processes obtain information from the environment for the higher-level representations that often require more effort to maintain. The feeling of 'flow' also is related to effortless attention; one's focus is on the mechanics of a physical activity with very little effort or attention to other forms of stimulation (Csikszentmihalyi 1997). Flow can be described as being closer to a more primary conscious experience and is often said to be an experience of being "in the moment," which does not include the reflective self-awareness more typical of higher-order thoughts. This feeling of flow may be experienced during extended periods of repetitious exercise like jogging, or when performing any sort of skilled activity that no longer requires a willful state of attention. For example, Lloyd, the writer and expert pianist from chapter 2, often experiences flow when writing novels or playing music.

Effortful attention, on the other hand, is described as focused, deliberate, voluntary, or goal-driven, and produces a subjective feeling of expending effort. This type of attention requires more resources to engage and maintain, since it requires a *focus* on the target object or task at hand. It often accompanies more complex tasks, whose maintenance calls for a top-down deployment of attention and working memory resources. Think about how it feels when you are first learning to drive a car or play the piano.

Some complex attentional processes, however, are so engrossing that they feel quite effortless, and one can lose a sense of time when performing such tasks (Bruya 2010). It is this version of effortless attention that may be particularly useful to study, since it is related to expertise and shows how memory systems can interact with attention to influence the perception of effort and time. The experience of driving home from work without remembering the exact process, which happens perhaps too often after many years of driving, may be an example. Effortless attention in this sense is a more controversial form of attention in relation to other forms of higher-level attentional processes, and has only recently begun to be discussed in the literature.

Assuming that effortless attention is a legitimate form of attention, it seems to have evolved recently and may be associated with the most experientially rich forms of conscious awareness. Necessarily, it would be a form of *conscious attention*. Moreover, it would also be one of the richest and most information-demanding forms of attention, suggesting a deep association with complex representation and more demanding cognitive tasks, even though these complex representations may not reach conscious awareness themselves. This type of effortless attention requires a level of expertise that develops after a repeated execution of willful, effortful attention (e.g., as implemented during procedural learning). It would be a form of attention related to learning capabilities of the most reflective, deliberate, and advanced type. Effortless attention also presupposes that the agent has self-awareness. Because of all these characteristics, it seems that effortless attention, including the experience of flow, must have evolved after many of the brain areas related to the more basic forms of attention developed into the form found in humans today. More important, it seems to demand the interrelation between conscious phenomenal awareness, or what it is like to experience oneself introspectively, and inputs from other forms of attention.

5.3 The Evolution of Conscious Attention and the Dissociation between Consciousness and Attention

So what does this survey of the various forms of conscious attention suggest about its evolutionary development? A brief sketch of the evolution of conscious attention may start, in chronological order, with: (1) selective attention for grounding features of the geometry of sensory space, which includes spatial and feature-based attention; (2) crossmodal detection of those features, which requires mappings across different feature maps as well as possible interfaces between different geometries; (3) conceptual content; (4) voluntary attention; and (5) forms of attention with very high cognitive demands but little experienced effort. How is this temporal pattern of the evolution of attention related to consciousness? Functional definitions of consciousness (e.g., access consciousness, high integration and complexity, or the global workspace theories discussed above) are only associated with items 3 through 5, and most accounts locate these capacities only in the very recent evolution of humans. Thus, a dissociation between consciousness and attention follows from strictly evolutionary considerations, *even if one can define consciousness functionally*—define it, that is, as the function that integrates information for global broadcast in a linguistic-like format.

The dissociation between consciousness and attention is even more severe if one considers conscious awareness to be a form of illusion (e.g., Humphrey 2011). Clearly, an extreme version of this view that considers consciousness to be an unsolvable mystery or a primitive feature of the universe that cannot be defined functionally will entail its full dissociation from attention. According to those views there is no possible overlap between attention, which is defined functionally, and consciousness, which cannot be defined functionally. Since we want to give an account of how conscious attention might have evolved, we shall focus on views that posit consciousness as illusory or a spandrel, but not on views that entail full dissociation with no possible overlap between the functions of attention and of phenomenal consciousness.

As we have said, anyone who subscribes to the view that consciousness transcends any functional explanation endorses the claim that considerations about evolution are irrelevant for the scientific study of consciousness, since evolutionary developments traditionally are explained in terms of functions. Some physicalist views on consciousness, however, *also* hold that evolutionary explanations are inadequate to account for the nature of consciousness, claiming that consciousness cannot be reduced to a set of functions subject to selection or adaptation processes (Block 1995). Furthermore, even those who adopt an evolutionary perspective in their explanation of consciousness reject the view that it is an adaptation (Carruthers 2000, Dennett 2005). Those who claim that consciousness has the purpose of giving us an illusory basis for social and empathic development (Humphrey 2011) seem to fall into the same category. Illusion accounts of consciousness, however, could also be understood as skeptical of the evolutionary explanation because the illusion of a unique subjective perspective may be considered pernicious rather than advantageous.

Nichols and Grantham (2000) argue, however, that the anatomical complexity of systems associated with phenomenal consciousness provides plenty of evidence that consciousness is in fact an evolutionary adaptation. Their primary claim is that phenomenally conscious experience involves inputs from different modalities as well as different channels within a modality, as for instance the different types of visual features that comprise feature-based attention. These various inputs must be integrated in some manner to produce a unified consciousness experience. The integration of these different channels, then, is an indication of a rather complex process and involves many structures. Therefore, the structural complexity that engenders phenomenal consciousness should support the idea that consciousness is an adaptation present in at least many vertebrates, as some philosophers, following Nagel (1974), assume.

The idea that consciousness serves the function of crossmodal and possibly intramodal integration has support from various theorists (for a review, see Baars 2002), but perhaps only in some novel circumstances (Mudrik, Faivre, and Koch, 2014). Notice, however, that based on our previous arguments, even views that characterize consciousness as adaptive would entail dissociation: crossmodal attention of such a highly integrative kind must have evolved much later than the basic forms of selective attention within modalities. Consciousness, even as an integrative mechanism, cannot be identical to attention because of the chronological dissociation between these early adaptations of low-level attentional systems and the more recently developed integrative forms of object-based and crossmodal attention.

Most extant accounts of the evolution of consciousness, then, seem to fit one of three types: (1) dualist accounts that reject any *physical* explanation of consciousness; (2) physicalist accounts that reject any *evolutionary* explanation of the nature of consciousness; and (3) physicalist accounts that propose evolutionary explanations of consciousness, but *deny that it is an adaptation* and claim that it is just a spandrel. All three options imply that conscious attention is not an adaptation because consciousness, unlike attention, is not adaptive. In other words, evidence of integration within and across attention systems that occurs without requiring conscious awareness (e.g., see Mudrik, Faivre, and Koch, 2014) reduces the likelihood that conscious attention is an evolutionarily motivated adaptation. Thus we confront the question of why, if the integration across systems is its purpose, would conscious awareness even exist if such integration occurs without it?

The three views just summarized entail that attention *cannot* be identical to consciousness and that conscious attention is the result of nonadaptive evolutionary processes. That is, most extant theories on the evolution of consciousness and attention require the conclusion that there is a strong dissociation between consciousness and attention. Furthermore, even views that consider consciousness as adaptive—the exception in the contemporary literature (e.g., Nichols and Grantham 2000)—require a dissociation between basic forms of attention and crossmodal integrative conscious attention.

Since the first two options consider evolution to be theoretically irrelevant in describing consciousness and conscious attention, we shall focus on the third option. There are at least two theories that fall within this third type. The first is that consciousness plays an important cognitive function and can thereby be defined in functionalist computational terms. According to this view, however, consciousness is not an adaptation but rather a spandrel, the result of an accidental combination among adaptive functions such as language and mind reading. Proponents of this view would argue that the main role of conscious awareness is to broadcast contents that are computed in a uniform format (presumably conceptual)—an essentially epistemic role since it provides *access* to contents across different modalities. Dennett (1969, 1991, 2005) and Carruthers (2000) defend this view, as do most theorists who favor 'global broadcast' or 'global workspace' views of consciousness (e.g., Baars 1988, Dehaene and Naccache 2001).

The second theory is that consciousness is a spandrel and that its function is not *epistemically driven*—unlike attention, which is epistemically driven because it provides access to contents that justify beliefs. Thus, according to this second view, consciousness is largely integrative, but serves an illusory function.[12] The thesis is that although consciousness is mainly an illusion, its function is to create the appearance of a coherent and unified unique self that enjoys private access to a world inaccessible to anyone else, with important ethical and spiritual consequences—a view defended by Humphrey (2011). In this sense, the Cartesian theater denounced by Dennett (1991) and previously by Gilbert Ryle (1949) is considered to be a positive achievement of the evolution of the human brain, in spite of its illusory character.

The main point of contention between these two spandrel views is whether the main purpose of consciousness is to broadcast information (and phenomenal experience is a spandrel) or to produce a useful (though admittedly unusual in terms of functionality) illusion associated with phenomenal content. We shall first demonstrate that although the broadcast view is the only one that relates consciousness and attention in a direct way in terms of access to contents, it still entails a severe dissociation between consciousness and attention. We then show that illusion views entail an even more severe form of dissociation.

An important source of confusion regarding the broadcast view is that it seems to suggest that consciousness must be identical to attention, because the purpose of attention—at least crossmodal attention—seems to be to broadcast and give access to information (see J. J. Prinz 2012). As we argued, however, this claim requires disambiguation because of the different functions and evolutionary development of distinct types of attention. The less dissociative interpretation of this access view is that selective attention is an early adaptation, while crossmodal attention to conscious contents is a later adaptation. The broadcast view clearly entails dissociation, challenging the identity between consciousness and attention. Moreover, crossmodal attention may not be sufficient for enabling a global *access* to uniformly formatted contents, although attention seems to be necessary for global access.

Even if crossmodal attention guaranteed access to uniformly formatted contents, it would still be a later evolutionary achievement. Such access would require the development of a global format, the broadcast and integration of information from different specialized areas of the brain, and the presence of all the basic forms of selective attention discussed above. These requirements strongly suggest a dissociation between basic forms of attention and conscious crossmodal attention. Related to this theoretical account of consciousness are the recent explanations of reentrant processing among brain structures (i.e., the feedback processing, in addition to feedforward, that appears important for complex attentional processes), which may be an indication of mechanisms that support consciousness (Di Lollo, Enns, and Rensink 2000, Hamker 2005, Lamme 2003, Seth and Baars 2005, Tononi and Koch 2008). Complex reentrant networks, especially with the frontal cortex, would be considered later adaptations.

An alternative theory, the field theory of consciousness, also proposes that contents from different modalities and brain regions (including those responsible for memory) are made conscious when they enter an integrative 'field of consciousness,' which is characterized by a phenomenal unity relation (see Bayne 2010, particularly chapter 1). Similarly, Bayne's (2007) theory of 'creature consciousness,' which specifies whether or not an organism can be said to be phenomenally conscious, requires two primary components: the integrative mechanisms that generate the 'phenomenal field' (of what it is like to be in a state, which may be related to activity in the thalamus); and mechanisms responsible for the specific contents of consciousness (e.g., neural inputs from the different cortical areas responsible for processing sensory information or memory-related thoughts). An analogous, but more emergentist, 'building block' theory proposes that each element of attention possesses its own 'micro-consciousness,' and these combine conjunctively (i.e., binding the different information via multi-stage integration) into a stable conscious awareness (Zeki and Bartels 1999). These proposals, however, require an *integration* of the different subsystems that is yet to be clearly explicated; such integration is unlikely to exist in the brain systems of more primitive organisms that exhibit signs of basic attention. Nevertheless, a unified consciousness is thought to appear when multimodal elements are somehow combined by some sort of integrating mechanism. Without the evolution of this mechanism, a unified phenomenal consciousness could not exist. When interpreted as a necessity claim, this evolutionary view on consciousness would entail a Type-A dissociation, but since such a view depends on a variety of functional definitions and evolution, the dissociation is best interpreted as Type-B.[13]

All other interpretations of spandrel views entail a stronger dissociation between consciousness and attention. In particular, illusion views characterize consciousness in terms of the unique and inaccessible perspective of a cohesive subjective self, which demands forms of introspection that are unlikely to be found across species and are certainly beyond the functional aspects of crossmodal attention. In other words, any view on consciousness that requires a subjective or phenomenal story limits how much of a relationship it could have with the various forms of attention found in humans and animals. Therefore, *all the possible* views on the evolution of consciousness and attention show that they *must* be dissociated.

Framing the relationship between consciousness and attention in adaptive terms helps solve a problem that tends to exacerbate the difficulty of studying conscious attention: many authors mean different things by the terms 'consciousness' and 'attention.' When authors use these terms, are they referring to what Block (1995) calls 'access' consciousness (the availability of representational content for use by various cognitive systems), or is it experience and the associated idea of phenomenality and subjectivity? Furthermore, is attention a combination of basic attention *plus* subjective experience (=conscious attention) or *just* attention (only selective information processing)? One crucial advantage of an evolutionary approach for understanding the relationship between consciousness and attention is that *it demonstrates that they must be dissociated regardless of how one defines them among the viable options currently available within the debate.* To the best of our knowledge, no other extant view on the relationship between consciousness and attention has this advantage.

Another important contribution of this evolutionary approach is that it shows that conscious attention cannot be either just simple selection or pure phenomenality. According to the functional views of consciousness that we have analyzed, voluntary and involuntary conscious attention play the role of broadcasting information that can be used for many cognitive purposes—information which is formatted in a common cognitive code. The formatting of conscious contents presents the following problem: whatever conscious attention is, it cannot be simple selection, or even crossmodal attention without quality access to highly integrated contents—a point that is uncontroversial. But then, conscious attention is a lot more than simply phenomenal consciousness because just experiencing contents (e.g., feeling pain or anxiety) need not have the selective and epistemic functions of attention. One can experience pain and pleasure without having any attention-driven process of selecting information and accessing contents. One also could have reflex-like reactions to stimuli and produce selective behavior that requires attention, but not experience things such as pain or pleasure. Furthermore, as has been confirmed empirically, many attentional processes can happen unconsciously (see section 2.3). This indicates that conscious attention is more than phenomenality and requires many subprocesses related to attention. We expand on this issue in section 5.4.

To summarize our arguments thus far, the earliest forms of attention, including those considered to be bottom-up (e.g., feature-based, spatial), act as selection mechanisms to filter relevant information for higher-level processing, such as the crossmodal mappings of different sensory information. These basic forms of attention have been identified across species and developed very early in the evolution of nervous systems. Through mechanisms such as feature maps and spatial indexing processes, a midlevel object-based attention evolved to facilitate the representation of complex multifeatured objects, to allow for the sustained maintenance of these representations in working memory structures, and to relate them to contents stored in long-term memory. This development may be the crucial step in the production of conscious attention, since these systems can perform more complicated tasks beyond simple feature detection and processing, including complex visual search tasks and object identification. Object-based attention, then, provides the scaffolding for more complex cognitive processes, including crossmodal integration; it also can maintain activation of representations (and consequently attention networks) and can facilitate communication and other social interactions (e.g., developing a theory of mind). These higher-level processes are heavily dependent on working memory systems to help in the maintenance and manipulation of this information. If a functionally motivated global workspace form of consciousness does indeed occur, it would be within the interaction between attention and memory, and here is where conscious attention would lie. Clearly, many basic forms of attention must happen before conscious attention occurs. Figure 5.1 illustrates these points.

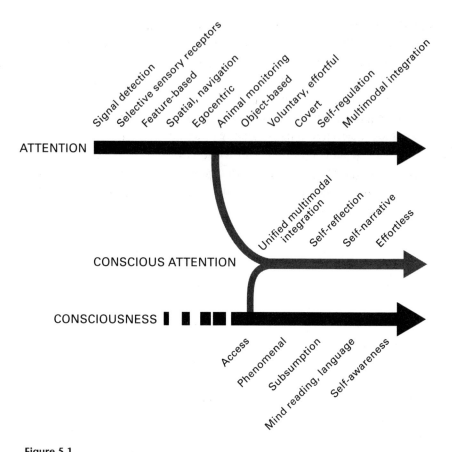

Figure 5.1
A rough sketch of the dissociation between consciousness and attention, and the possible overlap that may exist within conscious attention. The x-axis represents abstractly the temporal patterns of appearance for the different forms of consciousness and attention, but it does not imply a strictly linear progression. This overall sketch is generally uncontroversial for humans, since mounting evidence indicates that different neural systems serve consciousness and attention (e.g., Dehaene and Naccache 2001, Lamme 2006, Tallon-Baudry 2012, van Boxtel, Tsuchiya, and Koch 2010), and some theories of animal consciousness may imply a similar dissociation (e.g., Griffin and Speck 2004, Seth, Baars, and Edelman 2005). Our main proposal is that attention appeared relatively early, as an adaptation to serve various functional roles, and continued to evolve over time, whereas consciousness appeared later with no clear functional role. We must emphasize that proposals for a temporal order of appearance of the different forms of consciousness, however, are extremely controversial. Conscious attention, also a later development, may serve some adaptive functional purposes, but this is another controversial statement that awaits empirical support.

5.4 Conscious Attention—More Than a Feeling

Whatever consciousness is, it is certain that conscious attention cannot be a mere feeling or a pure 'what it is like' (as defined by Nagel 1974). Attention requires selection, short-term memory, representational contents, and the kind of cognitive processing that can be defined *functionally*. So whatever conscious attention is, it seems that it also should have a functional purpose. If conscious attention is indeed an integrative mechanism that combines inputs from different perceptual channels, as well as information stored in long-term memory, then we need to develop a better understanding of how the functions of integration are associated with subjective phenomenal experiences. What it is like to experience pain, for example—rather than simply responding to signals from pain receptors without the subjective experience—must be associated with the functional aspects of selective and crossmodal attention that are necessary to produce pain experiences—but many authors argue that it is impossible to give such an account. The challenge of understanding why conscious attention is not merely phenomenal experience highlights the difficulties ahead of us in providing a theory of conscious attention. Our main conclusion is that a definitive theory of conscious attention must incorporate the fact that evolutionary arguments indicate a dissociation between the basic forms of attention and conscious awareness.

In addition to the studies of blindsight mentioned previously (Brogaard 2011a, Kentridge 2012), further support for the dissociation between attentional processes and consciousness comes from studies on motor actions and conscious experience (Cohen et al. 2012, Desmurget et al. 2009, Kühn and Brass 2009, Wegner 2003). Such studies indicate that perceptual decisions and motor actions are executed before one becomes conscious of the decision or the intention to make the action, with some neural correlates supporting subjective experience and others supporting action (Filevich et al. 2013). These findings are often used to argue that consciousness is not necessary for producing actions, since there are background processes that execute these actions and we only "feel" like we are consciously doing things. Similarly, the separate ventral and dorsal pathways from the visual cortex indicate that information processing for determining what is seen ('vision for perception') can be independent from the execution of motor commands ('vision for action'), sometimes resulting in a dissociation between the information used to plan action and what enters conscious awareness. Under this account, the visual information that is processed to execute motor commands does not need to enter conscious experience—an idea that argues against the evolution of consciousness to serve an integrative function for performing actions. These considerations suggest, at the very least, that some forms of conscious attention may not be as globally formatted as some theories assume.

We shall now conclude with a more speculative hypothesis. We have shown that most of the extant views on the evolution of consciousness and attention entail a severe dissociation between consciousness and attention, and that even the few views that consider consciousness as adaptive also entail dissociation. We have further argued that this dissociation holds regardless of how 'attention' and 'consciousness' are defined and that *conscious attention* (the overlap between the evolutionarily distinct processes of consciousness and attention) must correspond to a strictly functional description of phenomenality. Here we will argue that if conscious attention exists, then it must be a *spectrum of different processes*, rather than a single type of process. Obviously, all the views we favor in this section reject full dissociation and allow for some form of overlap between consciousness and attention.

One source of data confirming a spectrum of conscious attention is the study of learning. Some researchers argue that the case of learning presents a dissociation between consciousness and attention. In particular, some claim that conscious awareness—but not attention—is necessary for learning (e.g., see Lovibond et al. 2011, Meuwese et al. 2013). Of course the problem is that the reportability of mental content *requires* consciousness, and the experiments typically implemented require reportability, either through verbal language or through gestures. Additionally, even when implicit perceptual learning is found under conditions where attention is supposedly not engaged (e.g., the inattentional condition in Meuwese et al. 2013), whether or not all forms of attention are truly absent in these cases remains debatable; we would argue that higher-level focused attention may be absent under these "inattentional" conditions, but low-level visual processing, such as feature-based attention, must still be operating. Furthermore, there certainly exist unconscious forms of learning, like procedural or associative learning in humans and animals. Such arguments indicate learning may involve a spectrum of conscious attention processes, some relying more heavily on phenomenal consciousness and others relying almost exclusively on low-level attention outside of phenomenal experience (corresponding to the effortful and effortless attention we described earlier). This relatively new research focus in relation to consciousness may be especially promising for revealing whether or not conscious attention indeed serves a functional purpose. In either case, it would still support a dissociative view.

Regardless of whether conscious attention is a single process or, as we believe, a spectrum of processes, the main conclusions of this book are as follows. First, evolutionary considerations show that attention and consciousness must be dissociated to some degree, ranging from functional dissociations (e.g., global versus more localized or within-modality attention) to severe dissociation (as most theories on the evolution of consciousness necessarily assume). This conclusion, moreover, entails that consciousness cannot be identical with attention. That is, although many attentional processes are necessary to provide the contents for conscious experience, attention and consciousness are not identical processes and cannot be reduced to one another given the range of dissociations we have discussed. Unlike any other extant view, our conclusions hold regardless of how the terms 'consciousness' and 'attention' are defined—as long as we agree that basic attention is the selective processing of information by low-level sensory mechanisms and that consciousness is the phenomenal experience that can be influenced by these processes. Finally, because consciousness and attention evolved separately, we cannot decisively conclude that conscious attention refers to a natural kind—this leaves an important open question that should be addressed with empirical methods.

It remains possible that the overlap between consciousness and attention, or conscious attention, serves a functional purpose, but this has yet to be clearly supported by empirical evidence, particularly in light of the integration that has been found to occur outside of conscious awareness. The task now is to examine this relationship, given that the most common forms of attention evolved independently and prior to any form of phenomenal consciousness. As we described in section 2.5, much research is happening on the experimental front that should lead to a richer understanding of consciousness. We believe that framing that work using the evolutionary considerations that we have presented will facilitate such research by promoting a focus on the dissociation that exists between consciousness and attention, and on how (or why) conscious attention seems to elude this dissociation. As mentioned previously, while most views on consciousness entail strong forms of dissociation, it is likely that at least *some* form of crossmodal attention is necessary for consciousness—another condition that has implications for a theory of conscious attention.[14]

5.5 Summary

Throughout this book we have argued that both empirical evidence and evolutionary considerations give us good reasons to believe that consciousness and attention are largely dissociated. There are some instances of overlap, as is the case with conscious attention, but overall, we must reject the identity view and any views that propose a full dissociation (although, full dissociation is more plausible than identity). In terms of our consciousness and attention dissociation (CAD) classification, we think there must be either a Type-A or Type-B CAD. Growing evidence suggests at least a Type-B CAD, which would account for several forms of attention that are not necessary for consciousness, with a single form of conscious attention. If there is a form of attention that is necessary for consciousness, and there is a single form of conscious attention (compatible with a Type-A CAD), then that kind of attention will very likely be a crossmodal form of attention, with high-quality semantic and epistemic access to contents. Notice, however, that even on this view there may be forms of attention that are not conscious.

Although there seem to be some forms of attention that are indeed necessary for consciousness (e.g., perceptual information must be processed in a selective way to provide the content of some forms of conscious experience), not *all forms* of attention are necessary. Additionally, some forms of conscious experience do not necessarily depend on the selection of information from the sense modalities or from long-term memory, such as the experience of pain or the phenomenological effects associated with autobiographical memory. In general, this conclusion is compatible with either Type-B or Type-C CAD.

By now, we hope you are convinced of the presence of a strong dissociation—or at least harboring doubts about the current theories on the relationship between consciousness and attention. We expect that the critical review of those theories presented here will help to advance our understanding of what remains one of philosophy and science's most puzzling problems: the nature of consciousness.

But *what is it?* you might still be wondering. We have skirted this issue because we still cannot say decisively what consciousness is, nor can we draw conclusions about its evolutionary purpose. Instead of promoting more speculation about the nature of consciousness, we have tried to point the way toward achieving a better understanding. So here is where we will present more speculative ideas.

What could be the purpose of consciousness? Based on our review of the empirical and philosophical literature, a few hypotheses become plausible. Let us say that conscious attention, and consciousness in general, is in fact an adaptation that promotes the survival of a species, ours in particular. How could we describe this adaptation? Perhaps its purpose is indeed to integrate cognitive processes and to provide a unified and stable phenomenal experience of the world; but in doing so, one of its main purposes could also be to *limit* the amount of information that enters into awareness in order to allow background processes to perform their critical functions smoothly. This proposal echoes our discussion of memory in section 4.2.2, where we noted that remembering all the information we encountered would be overwhelming and detract from having a memory that highlights more important events; we have evolved to best remember critical information, and our neurons tend to ignore a lot of detail (see Quiroga 2012). Let us elaborate on this *limited awareness* idea.

Based on experimental results in neuroscience and psychology indicating that a large part of our motor actions and decision processes occur prior to our being aware of them, it seems that consciousness has no clear functional or adaptive purpose. But, if the integrative, 'building block,' or 'subsumptive' views of consciousness are accurate descriptions, then it is possible that all the components of attention and other processes that happen in the background, allowing us to navigate through our environment, plan actions, and make decisions, possess a minimal phenomenal aspect that would be "on" by default. In simpler organisms, the combination of these phenomenal components in the presence of appropriate integration mechanisms would not be disorienting. For a complex organism such as humans, however, such a plethora of stimuli could result in an overwhelmingly chaotic experience that would be detrimental to survival: too much information would enter conscious experience. The result would perhaps be similar to some cognitive dysfunctions, such as extreme forms of schizophrenia.

Instead we could think of our unified but limited phenomenal experience as a mechanism that controls all the background noise so that the complex computations that occur in different areas of the brain can happen without mutual interference. In other words, humans have adapted to have a "functionally useless" and limited phenomenal consciousness in order to avoid the fragmented and chaotic state that we likely would otherwise experience. To use a familiar analogy, compare the music playing from your computer to a form of conscious experience for the computer itself. The auditory tones that compose the music are a result of all the background mechanics of the computer—the processor accessing the digital code in memory, the decoding and conversion of the digital files into the signals that are sent to the amplifier and speakers, and eventually the transformation of this data into wave frequencies that produce sound. But if all these background mechanics were audible in the foreground of the computer's "experience," the resulting state—the music—would be lost amid extraneous noise.

This analogy is limited, however, and it cannot adequately capture what makes conscious awareness so subjectively powerful. But music, as a product of complex sound-producing mechanisms, exemplifies an experience that is distinctively aesthetic and engaging. The aesthetic, empathic, and emotional strength characteristic of phenomenal consciousness is enabled partly by a similar process of stabilizing the contents of attention and reducing the amount of information that comes to the fore. Thus, consciousness does not just limit the background information in order to allow the subject to choose what to do. Crucially, the contents that do enter conscious awareness are infused with a force that cannot be explained semantically or epistemically. The function of consciousness is, in other words, to also make attentional contents urgently relevant and engaging.

Of course, this grand speculation is an idea in need of development, but if we had to assign a functional purpose to consciousness, it could possibly be this: to prevent all of the information processing related to attention and access consciousness from entering awareness, thus allowing for a smoother phenomenal experience that can immediately engage the subject. Some theories are compatible with this approach, such as field or building block theories, which assume that all our cognitive processes contain a little bit of consciousness, and that together these bits produce the full conscious experience that is familiar to us, in a manner analogous to the way energy accumulates in a field. Yet, even if our theory made sense and turned out to be robust, we would still be left with the questions of what *is* phenomenal experience and why does it even exist? So although this proposal is an idea worth developing, it does not contain a perfect answer to the 'why' of conscious experience. Essentially it argues that what is most distinctive about our limited phenomenal experiences is their *engaging* power, rather than their semantic content or epistemic characteristics.

This idea has important implications for effortless attention. If attention risks producing an overwhelming amount of phenomenal activity, the purpose of consciousness may be to produce an experience that constrains the information that is consciously attended, so that one doesn't phenomenally experience *all* perceptual aspects of the world, all the motor commands for implementing actions, or even the self, during effortless attention. In other words, the purpose of conscious attention here is not to reproduce reality, but to reduce the cognitive clutter that would impede the flow of experience. More important, since subjective experience is not a distinctive characteristic found in all forms of conscious attention, the engaging force of conscious awareness must be compatible with instances in which subjectivity is not experienced in any robust way.

Lloyd, our writer friend who is also an expert pianist, finds great joy in playing compositions he knows very well. This feeling of joy does not come from accessing all the information required to execute the movements to play the piano, but rather from engaging the process on a level that is effortless, even though it took a remarkable amount of training to be able to do so. Lloyd also experiences this feeling of effortlessness and flow when he composes music or writes long passages in his recent book; these activities, too, can be so engaging that he loses himself in the process. In other words, Lloyd no longer needs to be consciously aware of what he has to do in order to type on the computer keyboard or play music using the keys of a piano: the effortful stage of the process is obviated by his many years of training. These flow experiences exemplify how conscious attention may operate optimally, with minimal effort and a high level of engagement that allows the subject to be fully immersed in the present.

The importance of this feeling of timelessness and full engagement is not a new idea in the study of consciousness. It figures prominently in many areas of Eastern philosophy, as well as in Western conceptions of meditative practices. Although these perspectives cannot adequately be explored here, we should not ignore the ideas that emerge from them. For example, one way to think about consciousness from this perspective is that it is not an "acting" mechanism, that is, a mechanism geared toward performing actions. And yet, it is not entirely passive, either. Perhaps consciousness can best be thought of as a *mechanism for engagement*. When an emotion overwhelms the motor and cognitive systems, such as when extreme fear is provoked, the organism becomes immediately engaged in its environment in order to respond adequately; the survival circuits become active and take priority. This is a form of conscious attention, resulting from physiologic changes and based upon evolutionarily determined responses to a particular class of stimuli, which limits focus to one critical task.

Perhaps, then, consciousness evolved so that we can fully engage even without having to access all the other information that is being processed in the background, avoiding distractions that could be detrimental to the survival of a more cognitively complex organism. It may also serve the purpose of engaging the whole system in a powerful and meaningful way, so that we can not only react, report, or remember, but also empathize and understand the experiences of others. We could think of *attention* as a system for accessing information and interacting with the world, and of *consciousness* as a system responsible for engagement. Indeed, some authors suggest that the purpose of consciousness is just this, to be fully engaged with the self and the environment while not necessarily being in a state of higher-level self-reflection (e.g., see Humphrey 2011). Perhaps the adaptive purpose of consciousness is to provide a form of engagement with content which, instead of precisely reflecting reality, produces a limited but unified experience, reducing the conscious clutter to allow all the critical computations to occur in the background.

Regardless of whether or not consciousness serves an evolutionary purpose, we can still assert that it is not identical to attention. There are overlaps between the two mechanisms, as in the cases of conscious attention that we have described, but these cases may account for much less activity than what happens unconsciously; thus, consciousness and attention are largely dissociated. The main task now is to determine how they are related in the instances where they do occur in tandem; that is, to describe conscious attention empirically. Understanding this relationship between

CAD and the Evolution of Conscious Attention 215

consciousness and attention may help us solve the question of what is conscious experience by illuminating how it evolved.

One implication of our analysis of the dissociation between consciousness and attention is that theoretical and empirical approaches must focus particularly on the subset of cognitive phenomena associated with conscious attention. On the theoretical side, the almost universal emphasis on semantic content and epistemic features that dominates the literature on consciousness should be counterbalanced by an evaluation of the engagingly powerful *effects* of conscious content. One of the most important questions is how it is possible for conscious awareness to have such strong effects on motivation and perceptual engagement while at the same time being capable of unifying contents to provide remarkable stability and integrity. Our recommendation would be to study the type of cognitive enrichments that increase the degree of interpretive options regarding a given stimulus, as well as the aesthetic, moral, and emotional aspects of these enrichments. On the experimental side, it will be very important to locate the overlaps between consciousness and attention by contrasting findings on effortless flow experiences with findings on burdensome tasks that require effortful monitoring. That approach, based on the dissociation between consciousness and attention, should help to clarify some of the mysteries regarding conscious attention.

Notes

Chapter 1

1. A note on our use of quotation marks: 'single quotes' usually distinguish key theoretical terms; "double quotes" are reserved for direct quotations of text from another source, and are also used as so-called scare quotes, which indicate a more skeptical attitude toward the concept or terminology so enclosed (it will be obvious to the reader in which context double quotes are used).

2. We expand upon inverted spectrum cases in chapter 3. The basic idea is that two subjects may be looking at the same spectrum of light and have inverted color experiences. While the inverted spectrum is not an uncontroversial topic, we show that it is useful for understanding the relation between consciousness and attention.

3. It is possible to think of the orientations 'upward' and 'downward' as conceptual, but the point is that one may perceive upward and downward orientations without such concepts, while one may not be able to see a rabbit or a duck without the corresponding concepts. See chapter 4 for more on this issue.

4. Block (2010) defines 'mental paint' as the qualities of perception that are not captured by the representational content of stimuli or the external aspects of the stimuli one is directly aware of.

5. The terms 'recognitional' and 'phenomenal' self are used in Montemayor, Allen, and Morsella (2013). That paper also discusses 'mental paint,' but its main point is to report a new empirical finding regarding the less frequent changes in the phenomenology of the self when compared to changes experienced in the perception of ambiguous images.

6. These books include Carruthers (2000), Chalmers (1996), Dennett (2005), Koch (2004, 2012a), Lycan (1996), and Macphail (1998).

7. It is important to note that Varela's categorization is restricted to views on consciousness, defined loosely; obviously, for identity theories, this could also be a way to chart debates on attention.

Chapter 2

1. This idea is related to the notion of 'mental paint' introduced by Block (2010), which we discuss further in section 4.1.

2. Detailed descriptions of the organization of the visual cortex are available in Koch (2004, 49–86) and Hubel (1995, 93–125).

3. Addressing a similar topic are studies on the role of inference and probability in perception, which are crucial for understanding how perception generally works (see Clark 2013). Such findings indicate that scene statistics are an integral part of perceptual processes that are easily computed and can be used to optimize perception, suggesting an adaptive necessity for the systems that generate them.

4. Although we take such forms of attention that compute 'gist' as selective processing that can be used to perform actions and thus qualifies as 'attention,' some may disagree; see Wu (2014, 164–7) for a discussion.

5. Helmholtz was one of the first to report covert attention (Helmholtz 1866/1911).

6. We thank Mary Rorty for pointing out this connection.

7. This view is motivated by Henri Poincaré and his theoretical proposal of what it means to represent three-dimensional space in order to interact with objects in the environment and make a connection between mind and world; for example, by moving one's hand to keep a finger on a moving object (Poincaré 1913/1963, Pylyshyn 2007, 165–9).

8. For issues related to the idea of "divided attention," see Wright and Ward (2008, 53–60).

9. Incidentally, this finding plays a pivotal role as a motivation for Block's distinction between access consciousness and phenomenal consciousness, which entails a dissociation between consciousness and attention. We will discuss that in more detail in chapter 3.

10. Additionally, the *ideal observer model* can be useful in understanding how much information can be processed by comparing human performance to a theoretical ideal observer, which represents optimal information processing performance under given conditions (Geisler 1989, 2011). This model has been used in various areas of perception, including attention.

11. One could argue that learned rewards based on previous experience simply become part of the neural structures that modulate top-down attention, since the neural implementation of these associations can interact with attention. In any case, the main idea here is that the distinction between bottom-up and top-down attention may not be as clear-cut as originally thought.

12. A 'flow' experience is one that engages the subject in a way that produces a feeling of performing a task effortlessly—one that previously would have required more effortful processes (we will further discuss flow in sections 3.5 and 4.3).

13. One thing to note is that recent work regarding the specializations of brain areas has emphasized how neural *networks*, and the ability to form a distributed network of connections, developed, rather than focusing on how specific brain regions developed or on the size of these regions (Barton and Venditti 2013). This conceptualization suggests a different approach to the study of these systems and may be important for future work in this area. We will not address the related debate between connectionist (e.g., neural network) and symbolic or computational approaches in the study of the brain, but simply note that this debate is still active and will probably remain so.

14. This sort of lateral inhibition is necessary for the competitive neural network model, for example, as proposed by Itti and Koch (2001).

15. This perspective is already argued for furthering our understanding of memory by Randy Gallistel and Adam King (2009).

16. We cannot adequately discuss the issue of animal consciousness here, but see Boly et al. (2013), Feinberg and Mallatt (2013), and Griffin and Speck (2004).

17. An interesting proposal to consider is one by Marc Jeannerod, who suggests that "actions come into consciousness when perception does not match intention" (Jeannerod 2006). This could provide an explanation for why so many processes are completed outside of conscious attention and awareness, assuming that many actions are performed without such 'mismatches' (Bruya 2010, 16). Therefore, only the 'mismatches' tend to correspond with effortful attention and enter conscious awareness.

18. For a discussion on a related topic, see Newell and Shanks's (2014) review on how decision making may or may not be influenced by unconscious processes.

Chapter 3

1. Please refer to the glossary for specific definitions of consciousness and related terms.

2. This problem gets more complicated if one takes into consideration metaphysical views—for instance, whether or not phenomenal consciousness is irreducible to functional accounts—because the "hard problem" reflects a metaphysical rather than a purely epistemic gap. The metaphysical gap certainly entails the inadequacy of standard methods to study the mind, but so do several versions of the epistemic gap, including Block's phenomenalist approach, which we will discuss in more detail. Of the views discussed by Chalmers (2003a), type A materialism is the only

approach that clearly entails a functional account of consciousness and, thereby, a reduction of phenomenality to access.

3. Heterophenomenology, rather than introspective self-reports, is the specific methodology that Dennett (1991) promotes.

4. This, clearly, cannot be the standard sense of illusion in psychology and philosophy of perception, according to which the presence of misrepresented stimuli is a necessary condition for the illusion to occur. It is clearly not a hallucination either, since hallucinations concern specific stimuli. So, defining what sense of 'illusion' various authors have in mind is as puzzling as defining the phenomenon of consciousness itself.

5. The overflow argument should not be confused with the cognitive-overload argument in favor of a specific kind of higher-order theory of consciousness, which we explain below.

6. A disadvantage of this approach is that many subpersonal attentional mechanisms seem to have the defining features of attention; for example, they support demonstrative reference and are causally driven (see Pylyshyn 2007). We shall ignore this issue for the sake of conciseness and clarity—our main goal is to analyze the level of dissociation of extant views on the CAD spectrum.

7. Phenomenal conscious attention is associated with mental paint (Block 2010), which presents further problems (see section 4.1). Block, obviously, does not endorse these distinctions between types of attention, which we believe are compatible with the distinction between access and phenomenal consciousness. For empirical support concerning the richness of visual awareness and the limits of visual attention as a *non-illusory* phenomenon, see Vandenbroucke et al. (2014).

8. One may propose, for example, that representationism gets things generally right, but that there are a few odd cases, such as fringe feelings or appearance qualities, that seem representationally ineffable. One line of thought about the latter is that these can be explained as cases in which one is representing vaguely or imprecisely. Another possibility denies general representationism, but adopts a revised version of it that incorporates actions of an agent as constitutive of attended contents. For example, see Wu's (2011) impure reductive representationism.

9. Notice that the way in which the centering occurs must be elucidated. If the first-person indexical 'I' must be conceptualized to enable it, then linguistic capacities may be necessary for such centering, making the dissociation between attention and phenomenal consciousness rather severe. See Baker (2013) for such a proposal. For skepticism regarding the need for indexicals in thought and action, see Cappelen and Dever (2013).

10. J. J. Prinz (2012, 21–9), for instance, stipulates that consciousness does not require metarepresentation, and states as a desideratum for a successful theory of

consciousness that such a theory must be a first-order theory. Although we agree that first-order theories are more parsimonious and thus seem better suited to define phenomenal consciousness of the most basic form, the variety of relationships between different forms of attention and consciousness may require the kind of distinction higher-order theorists are interested in making.

11. Incidentally, as Byrne (2004) says, the distinction between conscious and unconscious experiences does not entail that first-order theories are false, since the main claim of such theories is that subtracting the metarepresentation that does the perspective embedding does not remove the phenomenal character of the experience (treated either representationally or phenomenally).

12. Our discussion has assumed that phenomenal consciousness is *state* rather than *creature* consciousness. Whether the property 'conscious' applies to mental states or to cognitive episodes is orthogonal to our discussion; the levels of dissociation apply equally to mental states and to events. If anything, the dissociation between attention and consciousness is more severe if one introduces self-awareness as a condition for consciousness.

13. There are two options about how to interpret this claim. One is that the richness of the phenomenology is illusory; although one feels that every time one targets an unconscious content it becomes available, in reality when a content is consciously available, that is the case simply because it is disposed to be globally accessible, without subjective probing or experiential richness. The other option is to accept that the phenomenology is indeed rich, but that one need not postulate tokenings of higher-order thoughts per content. These illusory and functional interpretations have important implications with respect to the evolution of consciousness and attention—an issue that we address in chapter 5.

14. See Block (1997) for these examples. Block claims that functionalists, like Dennett, identify very high quality access consciousness (in our example, Maria's) with phenomenal consciousness. Block, in contrast, says that access consciousness should be identified with medium access (in our example, Lucy's), in order to make the distinction between access and phenomenal consciousness as compelling as possible. This issue is deeply related to the overflow argument that Block uses in defense of the distinction between access and phenomenal consciousness, discussed above. It is worth mentioning that Thomas Reid may have anticipated this kind of case, in the context of strictly epistemic considerations. See Greco (2010, 35) and references therein.

15. There is room for debate about what exactly this means. Gareth Evans would say that consciousness is necessary for demonstrative *thought*, such as believing that a baseball is being pitched (Evans and McDowell 1982). But the notion of 'consciousness' involved here is functionally defined, and best understood as access consciousness. For a nonconceptual account of demonstrative thought that provides a representationalist theory of phenomenal content, see Michael Tye's (1995) PANIC

(poised, abstract, nonconceptual, intentional content) theory. The issue these views raise is whether or not the quality and degree of access boils down to the conceptual/nonconceptual content distinction. Regardless of how this issue is resolved, it seems that all forms of high-quality or lower-quality access to contents could be accommodated within the notion of 'access consciousness,' thus leaving the question of its dissociation from phenomenal consciousness open.

16. J. J. Prinz (2012) proposes such a distinction. A problem with introducing more distinctions is that it makes worse the already complicated set of ambiguities one finds in the literature on consciousness and attention. An advantage of our approach is that it demonstrates that consciousness must be dissociated from attention regardless of how they are defined, thereby avoiding *ad hoc* definitions and verbal disputes.

17. Uriah Kriegel (2013) argues that phenomenal intentionality is the source of other forms of intentionality and that at least part of what distinguishes it is how its contents are unified and determined, for instance, independently of any 'tracking' relations.

18. We analyze how subsumption relates to the self in section 3.4.

19. Tye (2003) offers a view that precludes the possibility of a parthood relation for experiences, partly based on temporal considerations: experiences cannot easily be quantified over time, and it seems better to characterize a whole temporal stretch of consciousness as a single experience. Dainton (2000) gives a simultaneous and a diachronic version of phenomenal unity, unlike Bayne (2010), who provides only a simultaneous version of it. Montemayor (2013) argues that there are two kinds of temporal integration, only one of which is metrically constrained.

20. See Baker (2013) for the view that language is necessary in order to have a robust first-person perspective.

21. If you have trouble imagining this, read Ashley Blocker's story, which describes how she deals with congenital insensitivity to pain (see Heckert 2012).

22. For discussion on mirror neurons and their relationship to consciousness, unconscious processing, and mind reading see Gallese and Goldman (1998) and Goldman (2006).

23. Some authors, including Baker (2013), disagree with this statement because they believe that all forms of self-awareness are self-recognitional and dependent on linguistic capacities for using indexicals. We believe this view is too strong. But we would like to emphasize that even such a view entails a dissociation between self-awareness and phenomenal awareness; so, regardless of who is right about self-awareness, CAD is entailed by those views as well.

24. We focus on virtue epistemology because of its emphasis on the agent's capacities and character traits.

25. Although the remainder of this section has some overlap with Montemayor (2014), that paper is not concerned with either attention or consciousness, and the material here is used only to motivate examples rather than provide arguments for dissociations between consciousness and attention.

26. There are similarities between Sosa's account of full knowledge and higher-order theories of consciousness, which we will not develop because they are not particularly insightful. What is central for our discussion here is how one can account for high-quality epistemic agency and responsibility without Sosa's reflective requirements in terms of access consciousness.

27. The classic findings on the unreliable nature of introspection by Nisbett and Wilson (1977) have inspired decades of voluminous research concerning unconscious biases. Findings concerning the limits of introspection also abound; for instance, with respect to blindsight, see Weiskrantz (2009). Maria, the super-duper-blindsighter, would accordingly be more reliable than subjects with phenomenal conscious introspections.

28. Evidence shows that unconscious syntax processing occurs reliably and automatically; see Batterink and Neville (2013).

29. See also Goodale (2011); for criticism of this finding, see de Brouwer et al. (2014).

30. Dualism or perhaps even panpsychism would be another alternative. But we will not consider these alternatives for three reasons. First, dualism complicates the theoretical considerations regarding the dissociation between access and phenomenal consciousness to a degree that goes beyond the empirically driven research that is required to adequately study such a dissociation. Second, the semantics for accuracy conditions concerning phenomenal concepts (in accordance with dualism) is controversial, and we do not need to endorse it at all to conclusively demonstrate degrees of dissociation between consciousness and attention. So we shall remain neutral with respect to the metaphysics and semantics of phenomenal consciousness. Third, all dualist views entail the dissociation of the functions of attention and the qualitative character of phenomenal experiences, so our arguments for CAD are reinforced by these views. Therefore, although we do not endorse dualist views, CAD is compatible with them.

31. We believe the best way to interpret this sense of 'mode of presentation' is Cath's (2009) notion of a *practical mode of presentation*, which he uses to defend the ability hypothesis. This issue, however, is not crucial to defending the normative claim we are making.

Chapter 4

1. We should note that our cognitive systems tend to do a good job of predicting content from a minimal number of features, as well as predictively allocating attention toward areas in the visual scene that are more likely to be informative (see Clark 2013).

2. For example, see Tversky (1977) and Goldstone (1994) for semantic approaches to similarity judgments and categorization. Applications of these semantic approaches also appear under the name of 'similarity measurement.' See also Kulvicki (2006) and Abell (2009) for philosophical views on resemblance. In scientific representation, resemblance is frequently associated with *isomorphism*—a bijective mapping—between members of two sets, for instance, from an empirical structure to a mathematical one and vice versa. See French (2003) and Giere (2004) for a characterization of the resemblance relation in terms of isomorphism. Dresner (2004) argues that measurements require only homomorphism, rather than isomorphism; that is, the empirical structure must be mapped into the mathematical one but not necessarily the other way around.

3. This is an example used by Fodor (2008) to illustrate the lack of principles of individuation in iconic representation.

4. Fodor (2008) says that the most distinctive feature of icons is that they lack the "items effect" that characterizes discursive symbols. By this, Fodor means exactly what Clark says in this passage: icons or pictures lack the required structure to count as discrete items because they lack categorical criteria for individuation.

5. For dissent regarding strict analogies between perception and visualization, see Pylyshyn (2003a).

6. Hans Reichenbach (1958) distinguishes an image-production capacity from the epistemic-normative capacity of geometry, which he associates with logic.

7. This distinction is introduced in Montemayor (in press), and some of the discussion in this section is based on that manuscript.

8. See Quiroga (2012) and references for evidence concerning neurons that ignore detail in order to create abstract patterns of thought.

9. These narrative effects can also be described in terms of first- and third-person perspectives. As mentioned previously, if one considers the first-person perspective to be dependent on either language or a primitive metaphysical property, then there will be a severe form of dissociation between autobiographical memory and other forms of memory, which will depend on the dissociation between self-aware conscious attention and other forms of conscious attention.

10. See Chu and Downes (2000) and references. Their experimental data specifically shows that odors are more effective cues for autobiographical memories than cues

Notes

from the other modalities, which demonstrates that in memory retrieval conscious attention does not operate in the same way or with the same evocative strength across modalities.

11. See Tsukiura and Cabeza (2011) for evidence concerning the shared neural correlates of aesthetic and moral judgment. See Zaidel and Nadal (2011) for a biological explanation of the evolution of the human traits underlying aesthetic and moral evaluation.

12. Similarly, one could imagine pain-free Ashley of the previous chapter complaining, "I want to *feel* pain, I don't want to be right about who is experiencing pain and how I should react to them." (A compelling aspect of her story is that Ashley really would like to feel pain so that it could guide her behavior and help her to understand others.) Or, more controversially, one can imagine an omniscient judge who knows everything about the fairest outcomes in human transactions but lacks the capacity for conscious experiences complaining: "I don't want merely to always be right about what is just; I want to experience hate, love, happiness, and *understand* others genuinely."

13. See Gertler (2011) for criticism of the transparency method and of similar psychologically based considerations.

14. With respect to the latter, we have in mind personal revelations in dreams like those reported by people who have lost a loved one and gain closure by dreaming about that person, realizing either implicitly (by how she looks) or explicitly (by what she says) that "everything is fine."

15. See Dunne (1927) for a fascinating discussion of dream retrieval, which includes guidelines for how to succeed in retrieving as much of the dream narrative as possible without confabulation. Dunne uses this technique to support his controversial account of time.

16. See Kahan and LaBerge (2011) and references. The authors confirmed that "dreaming and waking experiences are surprisingly similar in their cognitive and sensory qualities"—a verification of Dunne's (1927) less rigorous introspective findings. The difference is in narrative: dreams have a bizarre narrative. The evidence also challenges the thesis that no higher cognition happens in dreams, thus contradicting the imagery proposal.

17. See Wamsley and Stickgold (2010) for research that focuses on memory, and Mason et al. (2007) for research that focuses on stimulus-independent thought.

18. The relevant passages are in the Sixth Meditation (see Descartes 1904, 89–90).

19. Some may object that we have no knowledge of the phenomenology of our experiences (e.g., see Schwitzgebel 2011). We shall not address this skeptical worry here, but notice that it seems to assume a rejection of the continuity thesis based on an error theory about introspection, which also entails skepticism about autobiographical memory.

20. Lucid dreams create even more problems and comparisons, but note that there is evidence suggesting that neither bizarreness nor subjectively experienced realism of the dream seem to distinguish lucidity; see Voss et al. (2013).

21. Because of the emphasis on externally specified conditions for the constitution of traces, Halbwachs's view could be considered an 'active externalist' view on memory; see Sutton (2012) and references.

22. For the importance of the meaning of dreams and oneiric memory retrieval, see Freud (1923).

23. One thing to note about reality monitoring is that we may have an incorrect assessment as to the source of information (or the temporal order of events), but these memories still entail forms of conscious attention.

24. In support of our interpretation that these empathic cognitive enrichments give rise to a form of normativity that is not exclusively "rational," see Frazer (2010). Frazer claims that, in the context of the history of political science, this kind of empathic normativity underlies many ideals of the enlightenment.

Chapter 5

1. By 'reportable type of attention' we mean that the contents of attention, e.g., visual attention, are consciously accessible and experienced, such that one could report detecting the object of attention, be it a feature of an object or its spatiotemporal characteristics.

2. We presuppose throughout at least the minimal and uncontroversial definitions of consciousness as phenomenal experience, and of attention as the functional-selective neural mechanisms that enhance the processing of certain information in perception and cognition.

3. For a review of the underlying neural structures of attention and consciousness—which have implications about the possible evolutionary distinction between them—see Tallon-Baudry (2012).

4. Notice that the earlier evolution of forms of attention and the later evolution of conscious awareness is *compatible* with theories that claim that conscious awareness somehow emerges from, or is bootstrapped by, multiple attentional components: the emergence of different forms of conscious awareness would have occurred later, once the attentional components were in place. Notice also that this claim about evolution does not *entail* such emergentist claims, and is thus compatible with other views, even full dissociation views.

5. See also Corbetta et al. (1990), Corbetta and Shulman (2002), and Grent-'t-Jong and Woldorff (2007) for descriptions of the activations in frontal-parietal networks during the orientation of attention.

6. For a related review of this sort of analysis concerning the possible origins of consciousness in animals, see Feinberg and Mallatt (2013); for discussions on possible forms of a rudimentary 'theory of mind' in animals, see Horowitz (2011) and Udell, Dorey, and Wynne (2011).

7. By 'mind reading' we mean the cognitive capacity to identify and interpret the mental states of conspecifics.

8. As we discussed in section 2.1.1, this individuation of visual objects is related to demonstrative thoughts and proposes a solution to the 'reference problem,' which is concerned with how an object can be tracked through space and time and be linked to a mental representation (Perry 1997, Pylyshyn 2001, 2003b, 2007, Siegel 2002). From this perspective, attention anchors mental representations in the world by providing causal links to higher-level representations such as object files in visual working memory.

9. See McDowell (1994); for dissent concerning the need for concepts, see Stalnaker (1984).

10. For related reviews concerning the identification of consciousness in animals, see Edelman, Baars, and Seth (2005), Seth, Baars, and Edelman (2005), and Seth et al. (2008); for reviews on animal consciousness, see Griffin and Speck (2004), and Feinberg and Mallatt (2013).

11. The existence of self-awareness in animals remains debatable; see Baker (2013) for a skeptical view.

12. As mentioned previously, this sense of 'illusion' is not the standard one, in which an illusion is produced when a perceptual stimulus is misrepresented. The exact sense the authors have in mind when they use the term 'illusion' in this context requires clarification, but we do not delve into that issue here.

13. We should note that even under the 'field,' 'building block,' or other integrative theories of consciousness, we are still faced with the problem of identifying the purpose of the phenomenal aspects of conscious awareness that we as humans experience. Why should they exist discretely, eventually comprising our experience of consciousness? Similarly, what is it about the 'field of consciousness' that enables phenomenal experience?

14. Some of the arguments presented in this chapter were introduced in Haladjian and Montemayor (in press).

Glossary

access consciousness The type of consciousness associated with information that is available for thought, decision making, reporting, and action. Global workspace theories are about access consciousness, and it is controversial whether they suffice to explain phenomenal consciousness.

attention As James describes, attention is the ability to focus on "one out of what seem several simultaneously possible objects or trains of thought" and consequently "implies the withdrawal from some things in order to deal effectively with others" (James 1890/1905, 403–404). In other words, attention is composed of a number of mechanisms that selectively enhance, filter, and process information within the brain, often guided by the task at hand via top-down influences, or from salient features via bottom-up influences. Attention can be further broken down into the following varieties: feature-based, object-based, spatial, divided, focused, global, covert, and overt.

> ***bottom-up attention*** A stimulus-driven, exogenous, involuntary, effortless form of low-level attention; includes attentional capture, pop-out, visual indexing, and preattentive processes.
>
> ***conscious attention*** A phenomenal experience of the perceptual information processed by attentional mechanisms (which includes visual or auditory features) and thus available to cognition in a reportable manner. In a philosophical sense, conscious attention requires a demonstrative awareness of attending to a specific object (e.g., 'that' or 'this' object); it also entails voluntarily maintaining attention to an external object that has been perceptually selected (see Wu 2011).
>
> ***covert attention*** The ability to attend to something outside the center of eye gaze (in contrast to overt attention).
>
> ***divided attention*** Attention that is distributed among several objects or locations; although it remains debatable whether attention can truly be divided simultaneously and discretely among multiple objects or locations.

effortless attention A form of attention or action that (1) is not experienced as effortful or (2) does not involve exertion so that, due to the autotelicity of experience, subjective effort is lower than in normal conditions, while effectiveness is maintained at a normal or elevated level. *Objective effort* (exertion) is an increase in the metabolic or physiological processes of movement (physical effort) or thought (mental effort). *Subjective effort* is the feeling of exertion. Note that these can be dissociated and they do not map neatly. *Effortful* describes a form of attention in which there is subjective effort under normal conditions. *Autotelic* describes an experience in which one feels that the activity itself provides the impetus for continued action, involving a challenging activity that requires skill, the merging of behavior and awareness, clear goals and immediate feedback, concentration on the task at hand, a feeling of being in control, a loss of self-consciousness, and an altered sense of time. *Postvoluntary attention* (as defined in Bruya 2010, 5) is also a form of effortless attention in the second sense given above.

feature-based attention Attention to a particular kind of feature, such as color or orientation, which enhances the detection of this feature by modulation from specific sensory neurons.

focused attention Selective attention to a specific object, feature, or location.

global attention Attention to the overall or Gestalt characteristics of a scene, often referred to as 'gist' or as a distributed attention; tends to be coarser than focused attention.

object-based attention Here, the basic unit of attention is taken to be discrete 'objects' and the features that bind together to form those object-based representations.

spatial attention Attention to locations and not necessarily to specific objects or features.

top-down attention An endogenous, goal-oriented, voluntary, effortful form of focused attention; demonstrated, for example, when one performs a complex visual search task.

awareness A mental state, including perceptual experience, that is reportable; this is generally interchangeable with consciousness. See higher-order theories and 'what is it like.'

blindsight The inability to report perceiving a stimulus, combined with the ability to perform an action on it (e.g., reach for an object without being able to report seeing it), typically due to brain damage in the primary visual cortex. This ability to accurately identify aspects of a visual stimulus (e.g., location, type of movement) without conscious awareness of *any* stimuli is known as type 1 blindsight. Type 2 blindsight refers to instances when the person feels that something has changed but cannot identify the change as a visual percept.

Glossary

consciousness In general, consciousness refers to the ability and awareness required to report on an experience; it is a phenomenal experience related to qualia (and, for our purposes, agnostic to the distinction of 'brain state' versus 'mental state').

consciousness as 'what is it like' As defined by Nagel (1974), "No matter how the form may vary, the fact that an organism has conscious experience at all means, basically, that there is something it is like to be that organism. Something it is like for the organism." This is associated with the so-called *creature consciousness*, and is different from *mental-state consciousness*, that is, what makes a mental state a conscious one.

consciousness and the first-person perspective Consciousness of the phenomenal kind is associated with subjectivity as the vantage point from which conscious mental content is experienced. Nagel says: Every subjective phenomenon is essentially connected with a single point of view and this point of view is not describable in terms of 'third-person' explanations. The view that one needs self-awareness to be conscious is more controversial (see Koch 2012).

dualism (of the kind that Chalmers [2003a] calls type D) The microphysical world is not causally closed, and the explanatory gap is not only of an epistemic nature because there is also an ontological gap. This is the view called substance dualism or interactionism. (Proponent: Descartes)

dualism (type E) Phenomenal properties are distinct (ontologically) from physical properties, but have no effect on the physical world. This is the view called epiphenomenalism. (Proponent: Jackson, 1982)

epiphenomenal consciousness The type of consciousness proposed by Frank Jackson with the Knowledge argument. The proposal is that qualia are properties that are not reducible to physical properties. These properties are constitutive of consciousness and are causally irrelevant (see type E dualism).

evolutionary consciousness Consciousness as a capacity that evolved either accidentally or for a particular function. Integration theories seem to assume that consciousness has the function of unifying and integrating information for global manipulation. Other theories, which associate consciousness with higher-order thoughts (e.g., Carruthers) explain the evolution of consciousness as a consequence of the evolution of language and mind reading.

first-order representation (FOR) theories of consciousness FOR theories deny that higher-level representations are necessary for phenomenal consciousness. Thus, these theories maintain that having a first-order experience (having a standard perceptual experience with a specific representational content about the environment) suffices to produce phenomenal consciousness. According to HOR theories there is an important distinction between a mere experience and a conscious experience (many experiences, all the first-order level ones, are unconscious). For the FOR theorist, all experiences that have representational content are conscious. Note: FOR and HOR theories can be either phenomenist or intentionalist, so the latter distinction is independent from the former. (Proponents: Dretske, Tye, and Block)

Gestalt In relation to what is treated as a visual object, Gestalt principles include proximity, common fate, continuation, closure, and similarity; the term also describes features that are useful for organizing visual scenes and identifying individual objects.

"hard problem" There are easier problems of consciousness, such as how to explain it in terms of function, information processing, and neurology, and then there is the "hard problem" of consciousness, which is concerned with the question of why cognitive function is accompanied by phenomenal experience. The hard problem of consciousness, as defined by Chalmers (1996), is essentially the problem of subjective experience.

heterophenomenology Dennett's proposal for a third-person study of consciousness using standard empirical methods—that is, recording raw data and interpreting it—as a methodology for the scientific study of consciousness.

higher-order perception (HOP) theories of consciousness HOP theories, like HOT theories, propose that just having an experience is not sufficient to be conscious of that experience. Consciousness requires that an experience be experienced a certain way for an individual (the notion of 'what it is like'). For HOP theories, it means that the experience (for example, a first-order perceptual representation) must be experienced in the presence of a higher-level perceptual-like experience (or be embedded in a second-order perception): the perceptual experience that one has a first-order experience (consciousness requires a relationship among experiences, rather than a relationship between a first-order experience and the world). The higher-order perception is a necessary condition for phenomenal consciousness. (Proponents: Carruthers, Lycan, and Armstrong)

higher-order representation (HOR) theories of consciousness Both HOT and HOP are higher-order representation theories of phenomenal consciousness. (Access consciousness is supposed to be amenable to strictly functional accounts, but that idea is controversial, as is the relationship between HOT and HOP theories of phenomenal consciousness.)

higher-order thought (HOT) theories of consciousness HOT theories propose that just having an experience is not sufficient to be conscious of that experience. Consciousness requires that an experience be experienced a certain way for an individual (the notion of 'what is it like'). For HOT theories, it means that the experience (for example, a first-order perceptual representation) must be experienced in the presence of a thought (or be embedded in a thought): the thought that one is having the experience with the first-order content. These thoughts are supposed to be immediate and noninferential and are compatible with minimalist accounts of the self (they do not require dualism or Cartesian souls). The higher-order thought is a necessary condition for phenomenal consciousness. (Proponent: Rosenthal)

Glossary

inattentional blindness The inability to notice scene changes either because minor perceptual changes in the scene are hard to detect or because attentional (and working memory) resources are occupied by a simultaneous task that prevents attention from performing optimally.

intentionalism The view that the qualitative character of conscious experiences (their phenomenality) is entirely dependent on their representational content. The claim is that the phenomenal character of conscious experiences supervenes on their representational content, such that there cannot be differences with respect to phenomenal character without differences in representational content. (Proponents: Dretske and Tye)

materialism (type A) Any view that denies that there is an epistemic gap, holding that there is no "hard problem" of consciousness and there is no gap between physical-functional explanations and explanations of experiences. According to Chalmers's classification, these type-A views are typically reductive or functionalist. (Proponents: Dennett, Dretske, Rey, and Ryle)

materialism (type B) Any view that accepts there is an epistemic gap, but denies that the gap is ontological or metaphysical. Functional explanations do not solve the hard problem, but this does not imply an ontological distinction between consciousness and the physical world. (Proponents: Loar, Papineau, and Nagel, at least in some parts of his work)

materialism (type C) Same as type B, but with the additional claim that the gap is "closable." (Nagel also seems to defend this view.) Chalmers says that this is an unstable view that collapses into either type A or B materialism.

mental paint As defined by Block (2010), qualities of perception that are not captured by the representational content of stimuli or the external aspects of the stimuli one is directly aware of.

mind reading The cognitive capacity to identify and interpret the mental states of conspecifics.

monism (type F, nonreductive, typically not physicalist) Consciousness is constituted by the intrinsic properties of fundamental physical entities. This view also may be called panpsychism. Chalmers says that it may be the most plausible view about consciousness and that in general, dualist views are better (on scientific grounds) than materialist views.

mysterianism (physicalism without reduction; also called type B materialism by Chalmers) As defined by Nagel, the subjective (conscious) experience does not fit the standard pattern of explanation-as-reduction. The idea of moving from appearance to reality seems to make no sense with respect to this issue. This does not show that physicalism about consciousness is false, but it shows that we do not know how it could be true.

phenomenal consciousness The kind of consciousness that provides a qualitative character to experiences and that is not reducible to functional accounts; this is typically associated with the "hard problem."

phenomenal contrast Assumes that two experiences with two different perceptual contents should differ phenomenally as well; therefore, two experiences will contrast phenomenally because one has the hypothesized content while another experience will not. (Proponent: Siegel)

phenomenism The view that rejects intentionalism by rejecting the idea that phenomenal character supervenes on representational content. The claim is that representational content does not suffice to explain the phenomenal character of experiences. This view comes in degrees, from strong denials of representationism claiming that representationism is false in general to weak denials claiming the cases where it is false are few and isolated. Examples supporting phenomenism include the inverted spectrum. (Proponent: Block)

subsumption According to Bayne and Chalmers (2003), subsumption is the relation that unifies any set of conscious experiences that a subject has at a time into a cohesive overall experience, and provides a kind of unity that *cannot* be explained exclusively in terms of neurological, spatiotemporal, object-based, functional-representational, self-based, or rational relations.

unconscious According to some theories, conscious perception is all of the same kind, but unconscious perception is varied. One possibility is that there are multiple forms of attention for motor control that are all unconscious. This is further defined in terms of *subliminal* (a stimulus that does not cross the threshold for awareness) and *supraliminal* (a stimulus that is consciously perceived) effects.

References

Abell, Catharine. 2009. Canny resemblance. *Philosophical Review* 118 (2): 183–223.

Alais, David, John Cass, Robert P. O'Shea, and Randolph Blake. 2010. Visual sensitivity underlying changes in visual consciousness. *Current Biology* 20 (15): 1362–1367.

Allen, Allison K., Kevin Wilkins, Adam Gazzaley, and Ezequiel Morsella. 2013. Conscious thoughts from reflex-like processes: A new experimental paradigm for consciousness research. *Consciousness and Cognition* 22 (4): 1318–1331.

Allport, Alan. 1993. Attention and control: Have we been asking the wrong questions? A critical review of 25 years. In *Attention and Performance XIV*, ed. David E. Meyer and Sylvan Kornblum, 183–218. Cambridge, MA: MIT Press.

Alvarez, George A., and Patrick Cavanagh. 2004. The capacity of visual short-term memory is set both by visual information load and by number of objects. *Psychological Science* 15 (2): 106–111.

Alvarez, George A., and Aude Oliva. 2008. The representation of simple ensemble visual features outside the focus of attention. *Psychological Science* 19 (4): 392–398.

Alvarez, George A., and Aude Oliva. 2009. Spatial ensemble statistics are efficient codes that can be represented with reduced attention. *Proceedings of the National Academy of Sciences of the United States of America* 106 (18): 7345–7350.

Awh, Edward, Artem V. Belopolsky, and Jan Theeuwes. 2012. Top-down versus bottom-up attentional control: A failed theoretical dichotomy. *Trends in Cognitive Sciences* 16 (8): 437–443.

Awh, Edward, Harpreet Dhaliwal, Shauna Christensen, and Michi Matsukura. 2001. Evidence for two components of object-based selection. *Psychological Science* 12 (4): 329–334.

Awh, Edward, and John Jonides. 2001. Overlapping mechanisms of attention and spatial working memory. *Trends in Cognitive Sciences* 5 (3): 119–126.

Baars, Bernard J. 1988. *A Cognitive Theory of Consciousness*. Cambridge: Cambridge University Press.

Baars, Bernard J. 1998. The functions of consciousness. [Reply] *Trends in Neurosciences* 21 (5): 201.

Baars, Bernard J. 2002. The conscious access hypothesis: Origins and recent evidence. *Trends in Cognitive Sciences* 6 (1): 47–52.

Baars, Bernard J. 2005. Global workspace theory of consciousness: Toward a cognitive neuroscience of human experience. *Progress in Brain Research* 150:45–53.

Baars, Bernard J., William P. Banks, and James B. Newman. 2003. *Essential Sources in the Scientific Study of Consciousness*. Cambridge, MA: MIT Press.

Bacon, Elisabeth, Jean-Marie Danion, Francoise Kauffmann-Muller, and Agnès Bruant. 2001. Consciousness in schizophrenia: A metacognitive approach to semantic memory. *Consciousness and Cognition* 10 (4): 473–484.

Bacon, William F., and Howard E. Egeth. 1997. Goal-directed guidance of attention: Evidence from conjunctive visual search. *Journal of Experimental Psychology: Human Perception and Performance* 23 (4): 948–961.

Baddeley, Alan D. 2000. The episodic buffer: A new component of working memory? *Trends in Cognitive Sciences* 4 (11): 417–423.

Baddeley, Alan D. 2007. *Working Memory, Thought, and Action*. Oxford: Oxford University Press.

Baddeley, Alan D., and Sergio Della Sala. 1996. Working memory and executive control. *Philosophical Transactions of the Royal Society of London. Series B, Biological Sciences* 351 (1346): 1397–1493.

Baddeley, Alan D., and Lawrence Weiskrantz, eds. 1993. *Attention: Selection, Awareness, and Control: A Tribute to Donald Broadbent*. Oxford: Clarendon Press.

Baker, Lynne Rudder. 2013. *Naturalism and the First-person Perspective*. Oxford: Oxford University Press.

Ballard, Dana H., Mary M. Hayhoe, Polly K. Pook, and Rajesh P. N. Rao. 1997. Deictic codes for the embodiment of cognition. *Behavioral and Brain Sciences* 20 (4): 723–767.

Bar, Moshe, Roger B. Tootell, Daniel L. Schacter, Doug N. Greve, Bruce Fischl, Janine D. Mendola, Bruce R. Rosen, and Anders M. Dale. 2001. Cortical mechanisms specific to explicit visual object recognition. *Neuron* 29 (2): 529–535.

Bargh, John A., and Tanya L. Chartrand. 1999. The unbearable automaticity of being. *American Psychologist* 54 (7): 462–479.

Bargh, John A., and Melissa J. Ferguson. 2000. Beyond behaviorism: On the automaticity of higher mental processes. *Psychological Bulletin* 126 (6): 925–945.

Baron-Cohen, Simon. 1999. Can studies of autism teach us about consciousness of the physical and the mental? *Philosophical Explorations* 2 (3): 175–188.

Barton, Robert, and Chris Venditti. 2013. Human frontal lobes are not relatively large. *Proceedings of the National Academy of Sciences of the United States of America* 110 (22): 9001–9006.

Batson, Melissa A., Anton L. Beer, Aaron R. Seitz, and Takeo Watanabe. 2011. Spatial shifts of audio-visual interactions by perceptual learning are specific to the trained orientation and eye. *Seeing and Perceiving* 24 (6): 579–594.

Batterink, Laura, and Helen J. Neville. 2013. The human brain processes syntax in the absence of conscious awareness. *Journal of Neuroscience* 33 (19): 8528–8533.

Baylis, Gordon C., and Jon Driver. 1993. Visual attention and objects: Evidence for hierarchical coding of location. *Journal of Experimental Psychology: Human Perception and Performance* 19 (3): 451–470.

Bayne, Tim. 2007. Conscious states and conscious creatures: Explanation in the scientific study of consciousness. *Philosophical Perspectives* 21 (1): 1–22.

Bayne, Tim. 2010. *The Unity of Consciousness*. New York: Oxford University Press.

Bayne, Tim, and David J. Chalmers. 2003. What is the unity of consciousness? In *The Unity of Consciousness: Binding, Integration, and Dissociation*, ed. Axel Cleeremans, 23–58. Oxford: Oxford University Press.

Bays, Paul M., and Masud Husain. 2007. Spatial remapping of the visual world across saccades. *Neuroreport* 18 (12): 1207–1213.

Beauchamp, Michael S., Laurent Petit, Timothy M. Ellmore, John Ingeholm, and James V. Haxby. 2001. A parametric fMRI study of overt and covert shifts of visuospatial attention. *NeuroImage* 14 (2): 310–321.

Beaudot, William H. A., and Kathy T. Mullen. 2006. Orientation discrimination in human vision: Psychophysics and modeling. *Vision Research* 46 (1–2): 26–46.

Bermúdez, José Luis. 1995. Nonconceptual content: From perceptual experience to subpersonal computational states. *Mind & Language* 10 (4): 333–369.

Bernstein, Nicholai A. [1950] 1996. *Dexterity and Its Development*, ed. Mark L. Latash and Michael T. Turvey. Mahwah, NJ: L. Erlbaum Associates.

Blake, Randolph, Jan Brascamp, and David J. Heeger. 2014. Can binocular rivalry reveal neural correlates of consciousness? *Philosophical Transactions of the Royal Society of London. Series B, Biological Sciences* 369 (1641): 20130203.

Block, Ned. 1995. On a confusion about a function of consciousness. *Behavioral and Brain Sciences* 18 (2): 227–247.

Block, Ned. 1997. On a confusion about a function of consciousness. In *The Nature of Consciousness: Philosophical Debates*, ed. Ned Block, Owen J. Flanagan, and Güven Güzeldere, 375–415. Cambridge, MA: MIT Press.

Block, Ned. 2002. Concepts of consciousness. In *Philosophy of Mind: Classical and Contemporary Readings*, ed. David J. Chalmers, 206–218. New York: Oxford University Press.

Block, Ned. 2003. Consciousness. In *Encyclopedia of Cognitive Science*, ed. Lynn Nadel. New York: Nature Publishing Group.

Block, Ned. 2007. Consciousness, accessibility, and the mesh between psychology and neuroscience. *Behavioral and Brain Sciences* 30 (5–6): 481–499, discussion 499–548.

Block, Ned. 2010. Attention and mental paint. *Philosophical Issues* 20 (1): 23–63.

Block, Ned. 2011. Perceptual consciousness overflows cognitive access. *Trends in Cognitive Sciences* 15 (12): 567–575.

Boly, Melanie, Anil K. Seth, Melanie Wilke, Paul Ingmundson, Bernard Baars, Steven Laureys, David Edelman, and Naotsugu Tsuchiya. 2013. Consciousness in humans and non-human animals: Recent advances and future directions. *Frontiers in Psychology* 4 (625).

Bor, Daniel, and Anil K. Seth. 2012. Consciousness and the prefrontal parietal network: Insights from attention, working memory, and chunking. *Frontiers in Psychology* 3:63.

Botta, Fabiano, Juan Lupiáñez, and Ana B. Chica. 2014. When endogenous spatial attention improves conscious perception: Effects of alerting and bottom-up activation. *Consciousness and Cognition* 23:63–73.

Bradley, Francis Herbert. 1902. On active attention. *Mind* 11 (41): 1–30.

Brady, Timothy F., and Joshua B. Tenenbaum. 2013. A probabilistic model of visual working memory: Incorporating higher order regularities into working memory capacity estimates. *Psychological Review* 120 (1): 85–109.

Brefczynski, Julie A., and Edgar A. DeYoe. 1999. A physiological correlate of the 'spotlight' of visual attention. *Nature Neuroscience* 2 (4): 370–374.

Broadbent, Donald E. 1952. Listening to one of two synchronous messages. *Journal of Experimental Psychology* 44 (1): 51–55.

Broadbent, Donald E. 1958. *Perception and Communication*. New York: Pergamon Press.

References

Broadbent, Donald E. 1977. The hidden preattentive processes. *American Psychologist* 32 (2): 109–118.

Broadbent, Donald E., and Margaret H. P. Broadbent. 1987. From detection to identification: Response to multiple targets in rapid serial visual presentation. *Perception & Psychophysics* 42 (2): 105–113.

Broadbent, Donald E., and Margaret Gregory. 1963. Division of attention and the decision theory of signal detection. *Proceedings of the Royal Society of London. Series B, Biological Sciences* 158:222–231.

Brogaard, Berit. 2011a. Are there unconscious perceptual processes? *Consciousness and Cognition* 20 (2): 449–463.

Brogaard, Berit. 2011b. Centered worlds and the content of perception. In *A Companion to Relativism*, ed. Steven D. Hales, 137–158. Malden, MA: Wiley-Blackwell.

Brogaard, Berit. 2012. Non-visual consciousness and visual images in blindsight. *Consciousness and Cognition* 21 (1): 595–596.

Bruya, Brian. 2010. *Effortless Attention: A New Perspective in the Cognitive Science of Attention and Action*. Cambridge, MA: MIT Press.

Buckner, Randy L., and Fenna M. Krienen. 2013. The evolution of distributed association networks in the human brain. *Trends in Cognitive Sciences* 17 (2): 648–665.

Bundesen, Claus. 1990. A theory of visual attention. *Psychological Review* 97 (4): 523–547.

Bundesen, Claus. 2001. Attention: Models. In *International Encyclopedia of the Social & Behavioral Sciences*, ed. Neil J. Smelser and Paul B. Baltes, 878–884. Oxford: Pergamon.

Bundesen, Claus, Thomas Habekost, and Søren Kyllingsbaek. 2005. A neural theory of visual attention: Bridging cognition and neurophysiology. *Psychological Review* 112 (2): 291–328.

Burge, Tyler. 2010. *Origins of Objectivity*. Oxford: Oxford University Press.

Burkell, Jacquelyn A., and Zenon W. Pylyshyn. 1997. Searching through subsets: A test of the visual indexing hypothesis. *Spatial Vision* 11 (2): 225–258.

Burnham, Bryan R. 2007. Displaywide visual features associated with a search display's appearance can mediate attentional capture. *Psychonomic Bulletin & Review* 14 (3): 392–422.

Byrne, Alex. 2004. What phenomenal consciousness is like. In *Higher-order Theories of Consciousness*, ed. Rocco J. Gennaro, 203–226. Philadelphia, PA: John Benjamins Pub.

Byrne, Alex. 2005. Introspection. *Philosophical Topics* 33 (1): 79–104.

Calderone, Daniel J., Peter Lakatos, Pamela D. Butler, and F. Xavier Castellanos. 2014. Entrainment of neural oscillations as a modifiable substrate of attention. *Trends in Cognitive Sciences* 18 (6): 300–309.

Campbell, John. 1994. *Past, Space, and Self.* Cambridge, MA: MIT Press.

Campbell, John. 2002. *Reference and Consciousness.* New York: Oxford University Press.

Cappelen, Herman, and Joshu Dever. 2013. *The Inessential Indexical: On the Philosophical Insignificance of Perspective and the First Person.* 1st ed. Oxford: Oxford University Press.

Carey, Susan. 2001. Cognitive foundations of arithmetic: Evolution and ontogenesis. *Mind & Language* 16 (1): 37–55.

Carrasco, Marisa. 2009. Cross-modal attention enhances perceived contrast. *Proceedings of the National Academy of Sciences of the United States of America* 106 (52): 22039–22040.

Carrasco, Marisa. 2011. Visual attention: The past 25 years. *Vision Research* 51 (13): 1484–1525.

Carrasco, Marisa, Stuart Fuller, and Sam Ling. 2008. Transient attention does increase perceived contrast of suprathreshold stimuli: A reply to Prinzmetal, Long, and Leonhardt (2008). *Perception & Psychophysics* 70 (7): 1151–1164.

Carrasco, Marisa, Sam Ling, and Sarah Read. 2004. Attention alters appearance. *Nature Neuroscience* 7 (3): 308–313.

Carrasco, Marisa, and Yaffa Yeshurun. 2009. Covert attention effects on spatial resolution. *Progress in Brain Research* 176:65–86.

Carruthers, Peter. 2000. *Phenomenal Consciousness.* Cambridge: Cambridge University Press.

Carter, Olivia L., and Patrick Cavanagh. 2007. Onset rivalry: Brief presentation isolates an early independent phase of perceptual competition. *PLoS ONE* 2 (4): 1.

Carter, Olivia L., John D. Pettigrew, Felix Hasler, Guy M. Wallis, Guang B. Liu, Daniel Hell, and Franz X. Vollenweider. 2005. Modulating the rate and rhythmicity of perceptual rivalry alternations with the mixed 5-HT2A and 5-HT1A agonist psilocybin. *Neuropsychopharmacology* 30 (6): 1154–1162.

Cath, Yuri. 2009. The ability hypothesis and the new knowledge-how. *Noûs* 43 (1): 137–156.

Cavanagh, Patrick, Amelia R. Hunt, Arash Afraz, and Martin Rolfs. 2010. Visual stability based on remapping of attention pointers. *Trends in Cognitive Sciences* 14 (4): 147–153.

Cavanagh, Patrick, Angela T. Labianca, and Ian M. Thornton. 2001. Attention-based visual routines: Sprites. *Cognition* 80 (1–2): 47–60.

Cavanna, Andrea Eugenio, and Andrea Nani. 2008. Do consciousness and attention have shared neural correlates? *Psyche* 14 (1): 1–8.

Cave, Kyle R., and Narcisse P. Bichot. 1999. Visuospatial attention: Beyond a spotlight model. *Psychonomic Bulletin & Review* 6 (2): 204–223.

Chalmers, David J. 1995. Facing up to the problem of consciousness. *Journal of Consciousness Studies* 2 (3): 200–219.

Chalmers, David J. 1996. *The Conscious Mind: In Search of a Fundamental Theory*. New York: Oxford University Press.

Chalmers, David. 2003a. Consciousness and its place in nature. In *The Blackwell Guide to Philosophy of Mind*, ed. S. P. Stich and T. A. Warfield, 102–142. Malden, MA: Blackwell.

Chalmers, David J. 2003b. The nature of narrow content. *Philosophical Issues* 13 (1): 46–66.

Chen, Zhe. 2012. Object-based attention: A tutorial review. *Attention, Perception & Psychophysics* 74 (5): 784–802.

Cheries, Erik W., George E. Newman, Laurie R. Santos, and Brian J. Scholl. 2006. Units of visual individuation in rhesus macaques: Objects or unbound features? *Perception* 35 (8): 1057–1071.

Cherry, E. Colin. 1953. Some experiments on the recognition of speech, with one and with two ears. *Journal of the Acoustical Society of America* 25 (5): 975–979.

Chesney, Dana L., and Harry H. Haladjian. 2011. Evidence for a shared mechanism used in multiple-object tracking and subitizing. *Attention, Perception & Psychophysics* 73 (8): 2457–2480.

Chica, Ana B., Paolo Bartolomeo, and Juan Lupiáñez. 2013. Two cognitive and neural systems for endogenous and exogenous spatial attention. *Behavioural Brain Research* 237:107–123.

Choi, Hoon, and Takeo Watanabe. 2009. Selectiveness of the exposure-based perceptual learning: What to learn and what not to learn. *Learning & Perception* 1 (1): 89–98.

Chong, Sang Chul, and Anne Treisman. 2005a. Attentional spread in the statistical processing of visual displays. *Perception & Psychophysics* 67 (1): 1–13.

Chong, Sang Chul, and Anne Treisman. 2005b. Statistical processing: Computing the average size in perceptual groups. *Vision Research* 45 (7): 891–900.

Chou, Wei-Lun, and Su-Ling Yeh. 2012. Object-based attention occurs regardless of object awareness. *Psychonomic Bulletin & Review* 19 (2): 225–231.

Chu, Simon, and John J. Downes. 2000. Odour-evoked autobiographical memories: Psychological investigations of Proustian phenomena. *Chemical Senses* 25 (1): 111–116.

Chun, Marvin M., and René Marois. 2002. The dark side of visual attention. *Current Opinion in Neurobiology* 12 (2): 184–189.

Churchland, Patricia Smith. 1996. The Hornswoggle problem. *Journal of Consciousness Studies* 3 (5–6): 402–408.

Clark, Andy. 2013. Whatever next? Predictive brains, situated agents, and the future of cognitive science. *Behavioral and Brain Sciences* 36 (3): 181–204.

Clark, Austen. 2000. *A Theory of Sentience*. New York: Oxford University Press.

Cleeremans, Axel, and Luis Jiménez. 2002. Implicit learning and consciousness: A graded, dynamic perspective. In *Implicit Learning and Consciousness: An Empirical, Philosophical, and Computational Consensus in the Making*, ed. Robert M. French and Axel Cleeremans, 1–40. New York: Psychology Press.

Cohen, Elias H., and Frank Tong. 2013. Neural mechanisms of object-based attention. *Cerebral Cortex*. [Epub ahead of print]. 10.1093/cercor/bht303.

Cohen, Michael A., Patrick Cavanagh, Marvin M. Chun, and Ken Nakayama. 2012. The attentional requirements of consciousness. *Trends in Cognitive Sciences* 16 (8): 411–417.

Connor, Charles E., Howard E. Egeth, and Steven Yantis. 2004. Visual attention: Bottom-up versus top-down. *Current Biology* 14 (19): R850–R852.

Corbetta, Maurizio, Fran M. Miezin, Gordon L. Shulman, and Steven E. Petersen. 1991. Selective attention modulates extrastriate visual regions in humans during visual feature discrimination and recognition. *Ciba Foundation Symposium* 163:165–180.

Corbetta, Maurizio, Francis M. Miezin, Susan Dobmeyer, Gordon L. Shulman, and Steven E. Petersen. 1990. Attentional modulation of neural processing of shape, color, and velocity in humans. *Science* 248 (4962): 1556–1559.

Corbetta, Maurizio, and Gordon L. Shulman. 2002. Control of goal-directed and stimulus-driven attention in the brain. *Nature Reviews. Neuroscience* 3 (3): 201–215.

Cosmides, Leda, and John Tooby. 2013. Evolutionary psychology: New perspectives on cognition and motivation. *Annual Review of Psychology* 64:201–229.

Cowan, Nelson. 2001. The magical number 4 in short-term memory: A reconsideration of mental storage capacity. *Behavioral and Brain Sciences* 24 (1): 87–114; discussion 114–185.

Cowan, Nelson, Christopher L. Blume, and J. Scott Saults. 2012. Attention to attributes and objects in working memory. *Journal of Experimental Psychology: Learning, Memory, and Cognition* 39 (3): 731–747.

Crick, Francis, and Christof Koch. 1998. Consciousness and neuroscience. *Cerebral Cortex* 8 (2): 97–107.

Crick, Francis, and Christof Koch. 2003. A framework for consciousness. *Nature Neuroscience* 6 (2): 119–126.

Crick, Francis, and Christof Koch. 2005. What is the function of the claustrum? *Philosophical Transactions of the Royal Society of London. Series B, Biological Sciences* 360 (1458): 1271–1279.

Csikszentmihalyi, Mihaly. 1997. *Finding Flow: The Psychology of Engagement with Everyday Life*. 1st ed. New York: Basic Books.

Csikszentmihalyi, Mihaly, and Isabella Selega Csikszentmihalyi. 1988. *Optimal Experience: Psychological Studies of Flow in Consciousness*. Cambridge: Cambridge University Press.

Custers, Ruud. 2011. Disentangling attention and awareness: The case of predictive learning. *Consciousness and Cognition* 20 (2): 382–383.

Dainton, Barry. 2000. *Stream of Consciousness: Unity and Continuity in Conscious Experience*. New York: Routledge.

Damasio, Antonio R. 1994. *Descartes' Error: Emotion, Reason, and the Human Brain*. New York: G.P. Putnam.

Daprati, Elena, Nicolas Franck, Nicolas Georgieff, Joëlle Proust, Elisabeth Pacherie, Jean Daléry, and Marc Jeannerod. 1997. Looking for the agent: An investigation into consciousness of action and self-consciousness in schizophrenic patients. *Cognition* 65 (1): 71–86.

Dawkins, Richard. 2004. *The Ancestor's Tale: A Pilgrimage to the Dawn of Life*. London: Weidenfeld & Nicolson.

de Brouwer, Anouk J., Eli Brenner, W. Pieter Medendorp, and Jeroen B. J. Smeets. 2014. Time course of the effect of the Müller-Lyer illusion on saccades and perceptual judgments. *Journal of Vision* 14 (1): 4, 1–11.

de Fockert, Jan W., Geraint Rees, Chris D. Frith, and Nilli Lavie. 2004. Neural correlates of attentional capture in visual search. *Journal of Cognitive Neuroscience* 16 (5): 751–759.

de Gelder, Beatrice, Marco Tamietto, Geert van Boxtel, Rainer Goebel, Arash Sahraie, Jan van den Stock, Bernard M. C. Stienen, Lawrence Weiskrantz, and Alan Pegna. 2008. Intact navigation skills after bilateral loss of striate cortex. *Current Biology* 18 (24): R1128–R9.

de-Wit, Lee H., Geoff G. Cole, Robert W. Kentridge, and A. David Milner. 2011. The parallel representation of the objects selected by attention. *Journal of Vision* 11 (4):13, 1–10.

de-Wit, Lee H., Robert W. Kentridge, and A. David Milner. 2009. Object-based attention and visual area LO. *Neuropsychologia* 47 (6): 1483–1490.

Dehaene, Stanislas, Jean-Pierre Changeux, Lionel Naccache, Jérôme Sackur, and Claire Sergent. 2006. Conscious, preconscious, and subliminal processing: A testable taxonomy. *Trends in Cognitive Sciences* 10 (5): 204–211.

Dehaene, Stanislas, and Lionel Naccache. 2001. Towards a cognitive neuroscience of consciousness: Basic evidence and a workspace framework. *Cognition* 79 (1–2): 1–37.

Dehaene, Stanislas, Claire Sergent, and Jean-Pierre Changeux. 2003. A neuronal network model linking subjective reports and objective physiological data during conscious perception. *Proceedings of the National Academy of Sciences of the United States of America* 100 (14): 8520–8525.

Dennett, Daniel C. 1969. *Content and Consciousness*. London: Routledge.

Dennett, Daniel C. 1991. *Consciousness Explained*. 1st ed. Boston, MA: Little, Brown and Co.

Dennett, Daniel C. 2005. *Sweet Dreams: Philosophical Obstacles to a Science of Consciousness*. Cambridge, MA: MIT Press.

Descartes, René. [1904]. *Oeuvres de Descartes*. Edited by Charles Adam and Paul Tannery. 1904 ed. Vol. 7. Paris: J. Vrin.

Desimone, Robert, and John Duncan. 1995. Neural mechanisms of selective visual attention. *Annual Review of Neuroscience* 18:193–222.

Desmurget, Michel, Karen T. Reilly, Nathalie Richard, Alexandru Szathmari, Carmine Mottolese, and Angela Sirigu. 2009. Movement intention after parietal cortex stimulation in humans. *Science* 324 (5928): 811–813.

Dewart, Leslie. 1989. *Evolution and Consciousness: The Role of Speech in the Origin and Development of Human Nature*. Toronto: University of Toronto Press.

Di Lollo, Vincent, James T. Enns, and Ronald A. Rensink. 2000. Competition for consciousness among visual events: The psychophysics of reentrant visual processes. *Journal of Experimental Psychology. General* 129 (4): 481–507.

Donk, Mieke, and Jan Theeuwes. 2003. Prioritizing selection of new elements: Bottom-up versus top-down control. *Perception & Psychophysics* 65 (8): 1231–1242.

Dormashev, Yuri. 2010. Flow experience explained on the grounds of an activity approach to attention. In *Effortless Attention: A New Perspective in the Cognitive Science of Attention and Action*, ed. Brian Bruya, 287–333. Cambridge, MA: MIT Press.

Dorris, Michael C., Etienne Olivier, and Doug P. Munoz. 2007. Competitive integration of visual and preparatory signals in the superior colliculus during saccadic programming. *Journal of Neuroscience* 27 (19): 5053–5062.

Dresner, Eli. 2004. Over-assignment of structure. *Journal of Philosophical Logic* 33 (5): 467–480.

Driver, Jon. 2001. A selective review of selective attention research from the past century. *British Journal of Psychology* 92 (Part 1): 53–78.

Duncan, John. 1984. Selective attention and the organization of visual information. *Journal of Experimental Psychology. General* 113 (4): 501–517.

Duncan, John, and Glyn W. Humphreys. 1989. Visual search and stimulus similarity. *Psychological Review* 96 (3): 433–458.

Dunne, John William. [1927] 2001. *An Experiment with Time*. Charlottesville, VA: Hampton Roads Publishing Company.

Dux, Paul E., and Réne Marois. 2009. The attentional blink: A review of data and theory. *Attention, Perception & Psychophysics* 71 (8): 1683–1700.

Edelman, David B., Bernard J. Baars, and Anil K. Seth. 2005. Identifying hallmarks of consciousness in non-mammalian species. *Consciousness and Cognition* 14 (1): 169–187.

Egly, Robert, Jon Driver, and Robert D. Rafal. 1994. Shifting visual attention between objects and locations: Evidence from normal and parietal lesion subjects. *Journal of Experimental Psychology. General* 123 (2): 161–177.

Ehlers, Anke, Ann Hackmann, and Tanja Michael. 2004. Intrusive re-experiencing in post-traumatic stress disorder: Phenomenology, theory, and therapy. *Memory* 12 (4): 403–415.

Enns, James T., and Darlene A. Brodeur. 1989. A developmental study of covert orienting to peripheral visual cues. *Journal of Experimental Child Psychology* 48 (2): 171–189.

Eriksen, Charles W., and Yei-Yu Yeh. 1985. Allocation of attention in the visual field. *Journal of Experimental Psychology: Human Perception and Performance* 11 (5): 583–597.

Evans, Gareth, and John Henry McDowell. 1982. *The Varieties of Reference*. New York: Oxford University Press.

Fahrenfort, Johannes J., H. Steven Scholte, and Victor A. F. Lamme. 2007. Masking disrupts reentrant processing in human visual cortex. *Journal of Cognitive Neuroscience* 19 (9): 1488–1497.

Feigenson, Lisa, and Susan Carey. 2003. Tracking individuals via object-files: Evidence from infants' manual search. *Developmental Science* 6 (5): 568–584.

Feinberg, Todd E. 2011. The nested neural hierarchy and the self. *Consciousness and Cognition* 20 (1): 4–15.

Feinberg, Todd E., and Jon Mallatt. 2013. The evolutionary and genetic origins of consciousness in the Cambrian Period over 500 million years ago. *Frontiers in Psychology* 4: 667.

Feldman, Jacob. 2003. What is a visual object? *Trends in Cognitive Sciences* 7 (6): 252–256.

Feldman, Jacob. 2007. Formation of visual "objects" in the early computation of spatial relations. *Perception & Psychophysics* 69 (5): 816–827.

Feldman, Jacob, and Patrice D. Tremoulet. 2006. Individuation of visual objects over time. *Cognition* 99 (2): 131–165.

Ferster, David, and Kenneth D. Miller. 2000. Neural mechanisms of orientation selectivity in the visual cortex. *Annual Review of Neuroscience* 23:441–471.

Filevich, Elisa, Patricia Vanneste, Marcel Brass, Wim Fias, Patrick Haggard, and Simone Kühn. 2013. Brain correlates of subjective freedom of choice. *Consciousness and Cognition* 22 (4): 1271–1284.

Finlay, Barbara L., and P. B. Brodsky. 2006. Cortical evolution as the expression of a program for disproportionate growth and the proliferation of areas. In *Evolution of Nervous Systems: A Comprehensive Reference*, ed. Jon H. Kaas, 73–96. Amsterdam: Elsevier Academic Press.

Fitch, W. Tecumseh, Marc D. Hauser, and Noam Chomsky. 2005. The evolution of the language faculty: Clarifications and implications. *Cognition* 97 (2): 179–210, discussion 1–25.

Fodor, Jerry A. 1983. *The Modularity of Mind: An Essay on Faculty Psychology*. Cambridge, MA: MIT Press.

Fodor, Jerry A. 1998. *Concepts: Where Cognitive Science Went Wrong*. Oxford: Oxford University Press.

Fodor, Jerry A. 2008. *LOT 2: The Language of Thought Revisited*. New York: Oxford University Press.

Fougnie, Daryl, and René Marois. 2006. Distinct capacity limits for attention and working memory: Evidence from attentive tracking and visual working memory paradigms. *Psychological Science* 17 (6): 526–534.

Fox, Kieran C. R., Savannah Nijeboer, Elizaveta Solomonova, G. William Domhoff, and Kalina Christoff. 2013. Dreaming as mind wandering: Evidence from functional neuroimaging and first-person content reports. *Frontiers in Human Neuroscience* 7:412.

Franconeri, Steven L., George A. Alvarez, and Patrick Cavanagh. 2013. Flexible cognitive resources: Competitive content maps for attention and memory. *Trends in Cognitive Sciences* 17 (3): 134–141.

Franconeri, Steven L., George A. Alvarez, and James T. Enns. 2007. How many locations can be selected at once? *Journal of Experimental Psychology: Human Perception and Performance* 33 (5): 1003–1012.

Franconeri, Steven L., Andrew Hollingworth, and Daniel J. Simons. 2005. Do new objects capture attention? *Psychological Science* 16 (4): 275–281.

Franconeri, Steven L., and Daniel J. Simons. 2003. Moving and looming stimuli capture attention. *Perception & Psychophysics* 65 (7): 999–1010.

Frazer, Michael L. 2010. *The Enlightenment of Sympathy: Justice and the Moral Sentiments in the Eighteenth Century and Today.* New York: Oxford University Press.

French, Steven. 2003. A model-theoretic account of representation (or, I don't know much about Art—but I know it involves isomorphism). *Philosophy of Science* 70 (5): 1472–1483.

Freud, Sigmund. 1923. *The Interpretation of Dreams.* 3rd ed., ed. A. A. Brill. London: Allen & Unwin.

Friedenberg, Jay. 2013. *Visual Attention and Consciousness.* New York: Psychology Press.

Fukuda, Keisuke, Edward Awh, and Edward K. Vogel. 2010. Discrete capacity limits in visual working memory. *Current Opinion in Neurobiology* 20 (2): 177–182.

Gallese, Vittorio, and Alvin I. Goldman. 1998. Mirror neurons and the simulation theory of mind-reading. *Trends in Cognitive Sciences* 2 (12): 493–501.

Gallistel, Charles R. 1990a. *The Organization of Learning.* Cambridge, MA: MIT Press.

Gallistel, Charles R. 1990b. Representations in animal cognition: An introduction. *Cognition* 37 (1–2): 1–22.

Gallistel, Charles R. 1998. Brains as symbol processors: The case of insect navigation. In *Conceptual and Methodological Foundations*, ed. Saul Sternberg and Don L. Scarborough, 1–51. Cambridge, MA: MIT Press.

Gallistel, Charles R., and Adam Philip King. 2009. *Memory and the Computational Brain: Why Cognitive Science Will Transform Neuroscience.* Chichester: Wiley-Blackwell.

Gazzaley, Adam, and Anna C. Nobre. 2011. Top-down modulation: Bridging selective attention and working memory. *Trends in Cognitive Sciences* 16 (2): 129–135.

Gegenfurtner, Karl R. 2003. Cortical mechanisms of colour vision. *Nature Reviews Neuroscience* 4 (7): 563–572.

Gegenfurtner, Karl R., and Daniel C. Kiper. 2003. Color vision. *Annual Review of Neuroscience* 26:181–206.

Geisler, Wilson S. 1989. Sequential ideal-observer analysis of visual discriminations. *Psychological Review* 96 (2): 267–314.

Geisler, Wilson S. 2011. Contributions of ideal observer theory to vision research. *Vision Research* 51 (7): 771–781.

Geisler, Wilson S., Jeffrey S. Perry, Boaz J. Super, and Donald P. Gallogly. 2001. Edge co-occurrence in natural images predicts contour grouping performance. *Vision Research* 41 (6): 711–724.

Geng, Joy J. 2014. Attentional mechanisms of distractor suppression. *Current Directions in Psychological Science* 23 (2): 147–153.

Gertler, Brie. 2011. Self-knowledge and the transparency of belief. In *Self-knowledge*, ed. Anthony Hatzimoysis, 125–145. New York: Oxford University Press.

Giere, Ronald N. 2004. How models are used to represent reality. *Philosophy of Science* 71 (5): 742–752.

Goldman, Alvin I. 2006. *Simulating Minds: The Philosophy, Psychology, and Neuroscience of Mindreading*. New York: Oxford University Press.

Goldstone, Robert L. 1994. The role of similarity in categorization: Providing a groundwork. *Cognition* 52 (2): 125–157.

Goodale, Melvyn A. 1998. Vision for perception and vision for action in the primate brain. *Novartis Foundation Symposium* 218:21–34; discussion 34–39.

Goodale, Melvyn A. 2011. Transforming vision into action. *Vision Research* 51 (13): 1567–1587.

Goodale, Melvyn A., and A. David Milner. 1992. Separate visual pathways for perception and action. *Trends in Neurosciences* 15 (1): 20–25.

Gordon, Peter. 2009. Language and consciousness. In *Encyclopedia of Consciousness*, ed. William P. Banks, 447–459. Oxford: Academic Press.

Goujon, Annabelle, André Didierjean, and Sarah Poulet. 2014. The emergence of explicit knowledge from implicit learning. *Memory & Cognition* 42 (2): 225–236.

Gould, Stephen J., and Richard C. Lewontin. 1979. The spandrels of San Marco and the Panglossian paradigm: A critique of the adaptationist programme. *Proceedings of the Royal Society of London. Series B, Biological Sciences* 205 (1161): 581–598.

Gray, Jeffrey Alan. 2004. *Consciousness: Creeping Up on the Hard Problem*. New York: Oxford University Press.

Greco, John. 2004. How to preserve your virtue while losing your perspective. In *Ernest Sosa and His Critics*, ed. John Greco, 96–105. Malden, MA: Blackwell Publishing.

Greco, John. 2010. *Achieving Knowledge: A Virtue-Theoretic Account of Epistemic Normativity*. New York: Cambridge University Press.

Grent-'t-Jong, Tineke, and Marty G. Woldorff. 2007. Timing and sequence of brain activity in top-down control of visual-spatial attention. *PLoS Biology* 5 (1): e12.

Griffin, Donald R., and Gayle B. Speck. 2004. New evidence of animal consciousness. *Animal Cognition* 7 (1): 5–18.

Haenny, P. E., John H. R. Maunsell, and Peter H. Schiller. 1988. State dependent activity in monkey visual cortex. II. Retinal and extraretinal factors in V4. *Experimental Brain Research* 69 (2): 245–259.

Haladjian, Harry H., and Carlos Montemayor. In press. On the evolution of conscious attention. *Psychonomic Bulletin & Review*.

Haladjian, Harry H., Carlos Montemayor, and Zenon W. Pylyshyn. 2008. Segregating targets and nontargets in depth eliminates inhibition of nontargets in Multiple Object Tracking. *Visual Cognition* 16 (1): 107–110.

Haladjian, Harry H., and Zenon W. Pylyshyn. 2011. Enumerating by pointing to locations: A new method for measuring the numerosity of visual object representations. *Attention, Perception & Psychophysics* 73 (2): 303–308.

Haladjian, Harry H., Manish Singh, Zenon W. Pylyshyn, and Charles R. Gallistel. 2010. The encoding of spatial information during small-set enumeration. In *Proceedings of the 32nd Annual Conference of the Cognitive Science Society*, ed. Stellan Ohlsson and Richard Catrambone, 2839–44. Austin, TX: Cognitive Science Society.

Halbwachs, Maurice. [1952] 1992. *On Collective Memory*. Edited by Lewis A. Coser. Chicago: University of Chicago Press.

Hamker, Fred H. 2005. The reentry hypothesis: The putative interaction of the frontal eye field, ventrolateral prefrontal cortex, and areas V4, IT for attention and eye movement. *Cerebral Cortex* 15 (4): 431–447.

Hare, Caspar J. 2007. Self-bias, time-bias, and the metaphysics of self and time (Louis XIV). *Journal of Philosophy* 104 (7): 350–373.

Hare, Caspar J. 2009. *On Myself, and Other, Less Important Subjects*. Princeton, NJ: Princeton University Press.

Haugeland, John. 1998. *Having Thought: Essays in the Metaphysics of Mind*. Cambridge, MA: Harvard University Press.

Haxby, James V., Cheryl L. Grady, Barry Horwitz, Leslie G. Ungerleider, Mortimer Mishkin, Richard E. Carson, Peter Herscovitch, Mark B. Schapiro, and Stanley I.

Rapoport. 1991. Dissociation of object and spatial visual processing pathways in human extrastriate cortex. *Proceedings of the National Academy of Sciences of the United States of America* 88 (5): 1621–1625.

He, Sheng, Patrick Cavanagh, and James Intriligator. 1996. Attentional resolution and the locus of visual awareness. *Nature* 383 (6598): 334–337.

He, Xun, Glyn Humphreys, Silu Fan, Lin Chen, and Shihui Han. 2008. Differentiating spatial and object-based effects on attention: An event-related brain potential study with peripheral cueing. *Brain Research* 1245:116–125.

Heckert, Justin. 2012. The hazards of growing up painlessly. *The New York Times Magazine*, November 15. http://www.nytimes.com/2012/11/18/magazine/ashlyn-blocker-feels-no-pain.html.

von Helmholtz, Hermann. [1866] 1911. *Treatise on Physiological Optics*. 3 vols. Rochester, NY: Continuum.

Hickey, Clayton, John J. McDonald, and Jan Theeuwes. 2006. Electrophysiological evidence of the capture of visual attention. *Journal of Cognitive Neuroscience* 18 (4): 604–613.

Hoerl, Christoph. 1999. Memory, amnesia and the past. *Mind & Language* 14 (2): 227–251.

Hoffman, James E., and Baskaran Subramaniam. 1995. The role of visual attention in saccadic eye movements. *Perception & Psychophysics* 57 (6): 787–795.

Holcombe, Alex O. 2010. The binding problem. In *Encyclopedia of Perception*, ed. E. Bruce Goldstein, 205–208. Thousand Oaks, CA: SAGE Publications, Inc.

Holcombe, Alex O., and Patrick Cavanagh. 2001. Early binding of feature pairs for visual perception. *Nature Neuroscience* 4 (2): 127–128.

Holcombe, Alex O., and Wei-Ying Chen. 2012. Exhausting attentional tracking resources with a single fast-moving object. *Cognition* 123 (2): 218–228.

Hollingworth, Andrew, and Ashleigh M. Maxcey-Richard. 2013. Selective maintenance in visual working memory does not require sustained visual attention. *Journal of Experimental Psychology: Human Perception and Performance* 39 (4): 1047–1058.

Hommel, Bernhard. 2004. Event files: Feature binding in and across perception and action. *Trends in Cognitive Sciences* 8 (11): 494–500.

Hommel, Bernhard. 2005. How much attention does an event file need? *Journal of Experimental Psychology: Human Perception and Performance* 31 (5): 1067–1082.

Hommel, Bernhard. 2007a. Consciousness and control: Not identical twins. *Journal of Consciousness Studies* 14 (1–2): 155–176.

Hommel, Bernhard. 2007b. Feature integration across perception and action: Event files affect response choice. *Psychological Research* 71 (1): 42–63.

Hommel, Bernhard. 2010. Grounding attention in action control: The intentional control of selection. In *Effortless Attention: A New Perspective in the Cognitive Science of Attention and Action*, ed. Brian Bruya, 121–140. Cambridge, MA: MIT Press.

Hommel, Bernhard, and Lorenza S. Colzato. 2004. Visual attention and the temporal dynamics of feature integration. *Visual Cognition* 11 (4): 483–521.

Hommel, Bernhard, Klaus Kessler, Frank Schmitz, Joachim Gross, Elkan Akyürek, Kimron Shapiro, and Alfons Schnitzler. 2006. How the brain blinks: Towards a neurocognitive model of the attentional blink. *Psychological Research* 70 (6): 425–435.

Horowitz, Alexandra. 2011. Theory of mind in dogs? Examining method and concept. *Learning & Behavior* 39 (4): 314–317.

Horowitz, Todd S., Alex O. Holcombe, Jeremy M. Wolfe, Helga C. Arsenio, and Jennifer S. DiMase. 2004. Attentional pursuit is faster than attentional saccade. *Journal of Vision* 4 (7): 585–603.

Hsu, Shen-Mou, Nathalie George, Valentin Wyart, and Catherine Tallon-Baudry. 2011. Voluntary and involuntary spatial attentions interact differently with awareness. *Neuropsychologia* 49 (9): 2465–2474.

Hubel, David H. 1995. *Eye, Brain, and Vision*. New York: Scientific American Library.

Hubel, David H., and Torsten N. Wiesel. 1959. Receptive fields of single neurones in the cat's striate cortex. *Journal of Physiology* 148:574–591.

Hubel, David H., and Torsten N. Wiesel. 1962. Receptive fields, binocular interaction and functional architecture in the cat's visual cortex. *Journal of Physiology* 160:106–154.

Humphrey, Nicholas. 2011. *Soul Dust: The Magic of Consciousness*. Princeton: Princeton University Press.

Intriligator, James, and Patrick Cavanagh. 2001. The spatial resolution of visual attention. *Cognitive Psychology* 43 (3): 171–216.

Itti, Laurent, and Christof Koch. 2000. A saliency-based search mechanism for overt and covert shifts of visual attention. *Vision Research* 40 (10–12): 1489–1506.

Itti, Laurent, and Christof Koch. 2001. Computational modelling of visual attention. *Nature Reviews. Neuroscience* 2 (3): 194–203.

Jackson, Frank. 1982. Epiphenomenal qualia. *Philosophical Quarterly* 32:127–136.

James, William. [1890] 1905. *The Principles of Psychology*. New York: H. Holt & Company.

Jeannerod, Marc. 2006. *Motor Cognition: What Actions Tell the Self.* New York: Oxford University Press.

Jensen, Ole, Mathilde Bonnefond, and Rufin VanRullen. 2012. An oscillatory mechanism for prioritizing salient unattended stimuli. *Trends in Cognitive Sciences* 16 (4): 200–206.

Johnson, Jeffrey S., Andrew Hollingworth, and Steven J. Luck. 2008. The role of attention in the maintenance of feature bindings in visual short-term memory. *Journal of Experimental Psychology: Human Perception and Performance* 34 (1): 41–55.

Johnson, Marcia K. 1991. Reality monitoring: Evidence from confabulation in organic brain disease patients. In *Awareness of Deficit after Brain Injury: Clinical and Theoretical Issues*, ed. George P. Prigatano and Daniel L. Schacter, 176–197. New York: Oxford University Press.

Johnson, Marcia K., Shahin Hashtroudi, and D. Stephen Lindsay. 1993. Source monitoring. *Psychological Bulletin* 114 (1): 3–28.

Johnston, William A., and Veronica J. Dark. 1986. Selective attention. *Annual Review of Psychology* 37:43–75.

Jonides, John. 1983. Further toward a model of the mind's eye's movement. *Bulletin of the Psychonomic Society* 21 (4): 247–250.

Jonides, John, and Steven Yantis. 1988. Uniqueness of abrupt visual onset in capturing attention. *Perception & Psychophysics* 43 (4): 346–354.

Kahan, Tracey L., and Stephen P. LaBerge. 2011. Dreaming and waking: Similarities and differences revisited. *Consciousness and Cognition* 20 (3): 494–514.

Kahneman, Daniel. 1973. *Attention and Effort.* Englewood Cliffs, NJ: Prentice-Hall.

Kahneman, Daniel, Anne Treisman, and Brian J. Gibbs. 1992. The reviewing of object files: Object-specific integration of information. *Cognitive Psychology* 24 (2): 175–219.

Kant, Immanuel. 1929. *Critique of Pure Reason.* Smith, Norman Kemp (trans.). London: Macmillan.

Kastner, Sabine, and Mark A. Pinsk. 2004. Visual attention as a multilevel selection process. *Cognitive, Affective & Behavioral Neuroscience* 4 (4): 483–500.

Kastner, Sabine, and Leslie G. Ungerleider. 2000. Mechanisms of visual attention in the human cortex. *Annual Review of Neuroscience* 23:315–341.

Kastner, Sabine, and Leslie G. Ungerleider. 2001. The neural basis of biased competition in human visual cortex. *Neuropsychologia* 39 (12): 1263–1276.

References

Kentridge, Robert W. 2011. Attention without awareness: A brief review. In *Attention: Philosophical and Psychological Essays,* ed. Christopher Mole, Declan Smithies, and Wayne Wu, 228–246. Oxford: Oxford University Press.

Kentridge, Robert W. 2012. Blindsight: Spontaneous scanning of complex scenes. *Current Biology* 22 (15): R605–R606.

Kentridge, Robert W., Tanja C. W. Nijboer, and Charles A. Heywood. 2008. Attended but unseen: Visual attention is not sufficient for visual awareness. *Neuropsychologia* 46 (3): 864–869.

Kiefer, Markus. 2007. Top-down modulation of unconscious 'automatic' processes: A gating framework. *Advances in Cognitive Psychology* 3 (1–2): 289–306.

Klimesch, Wolfgang. 2012. Alpha-band oscillations, attention, and controlled access to stored information. *Trends in Cognitive Sciences* 16 (12): 606–617.

Knudsen, Eric I. 2007. Fundamental components of attention. *Annual Review of Neuroscience* 30:57–78.

Koch, Christof. 2004. *The Quest for Consciousness: A Neurobiological Approach.* Denver, CO: Roberts & Company Publishers.

Koch, Christof. 2012a. *Consciousness: Confessions of a Romantic Reductionist.* Cambridge, MA: MIT Press.

Koch, Christof. 2012b. Consciousness: We're getting very close. *New Scientist* 214 (2860): 24–25.

Koch, Christof, and Naotsugu Tsuchiya. 2007. Attention and consciousness: Two distinct brain processes. *Trends in Cognitive Sciences* 11 (1): 16–22.

Koch, Christof, and Naotsugu Tsuchiya. 2012. Attention and consciousness: Related yet different. *Trends in Cognitive Sciences* 16 (2): 103–105.

Koch, Christof, and Shimon Ullman. 1985. Shifts in selective visual attention: Towards the underlying neural circuitry. *Human Neurobiology* 4 (4): 219–227.

Kornblith, Hilary. 2010. What reflective endorsement cannot do. *Philosophy and Phenomenological Research* 80 (1): 1–19.

Kosslyn, Stephen M. 1994. *Image and Brain: The Resolution of the Imagery Debate.* Cambridge, MA: MIT Press.

Kouider, Sid, Vincent de Gardelle, Jerome Sackur, and Emmanuel Dupoux. 2010. How rich is consciousness? The partial awareness hypothesis. *Trends in Cognitive Sciences* 14 (7): 301–307.

Kovács, Ilona. 1996. Gestalten of today: Early processing of visual contours and surfaces. *Behavioural Brain Research* 82 (1): 1–11.

Krantz, David H. 1969. Threshold theories of signal detection. *Psychological Review* 76 (3): 308–324.

Kriegel, Uriah. 2013. *Phenomenal Intentionality*. New York: Oxford University Press.

Kühn, Simone, and Marcel Brass. 2009. Retrospective construction of the judgement of free choice. *Consciousness and Cognition* 18 (1): 12–21.

Kulvicki, John V. 2006. *On Images: Their Structure and Content*. New York: Oxford University Press.

LaBerge, David. 1983. Spatial extent of attention to letters and words. *Journal of Experimental Psychology: Human Perception and Performance* 9 (3): 371–379.

LaBerge, David. 1997. Attention, awareness, and the triangular circuit. *Consciousness and Cognition* 6 (2–3): 149–181.

LaBerge, David. 2001. Attention, consciousness, and electrical wave activity within the cortical column. *International Journal of Psychophysiology* 43 (1): 5–24.

LaBerge, David. 2002. Attentional control: Brief and prolonged. *Psychological Research* 66 (4): 220–233.

LaBerge, David. 2005. Sustained attention and apical dendrite activity in recurrent circuits. *Brain Research* 50 (1): 86–99.

LaBerge, David, Vincent Brown, Marc Carter, David Bash, and Alan A. Hartley. 1991. Reducing the effects of adjacent distractors by narrowing attention. *Journal of Experimental Psychology: Human Perception and Performance* 17 (1): 65–76.

Lakoff, George, and Mark Johnson. 1980. *Metaphors We Live By*. Chicago: University of Chicago Press.

Lamme, Victor A. F. 2003. Why visual attention and awareness are different. *Trends in Cognitive Sciences* 7 (1): 12–18.

Lamme, Victor A. F. 2004. Separate neural definitions of visual consciousness and visual attention; a case for phenomenal awareness. *Neural Networks* 17 (5–6): 861–872.

Lamme, Victor A. F. 2006. Towards a true neural stance on consciousness. *Trends in Cognitive Sciences* 10 (11): 494–501.

Lau, Hakwan C., Robert D. Rogers, Patrick Haggard, and Richard E. Passingham. 2004. Attention to intention. *Science* 303 (5661): 1208–1210.

Lavie, Nilli, Diane M. Beck, and Nikos Konstantinou. 2014. Blinded by the load: Attention, awareness and the role of perceptual load. *Philosophical Transactions of the Royal Society of London. Series B, Biological Sciences* 369 (1641): 20130205.

Leber, Andrew B., and Howard E. Egeth. 2006. It's under control: Top-down search strategies can override attentional capture. *Psychonomic Bulletin & Review* 13 (1): 132–138.

LeDoux, Joseph. 2012. Rethinking the emotional brain. *Neuron* 73 (4): 653–676.

Lee, Daeyeol, and Marvin M. Chun. 2001. What are the units of visual short-term memory, objects or spatial locations? *Perception & Psychophysics* 63 (2): 253–257.

Leontiev, Aleksei N. 1978. *Activity, Consciousness, and Personality*. Hall, Marie J. (trans.). Englewood Cliffs, NJ: Prentice-Hall.

Lewis, David K. 1969. *Convention: A Philosophical Study*. Cambridge, MA: Harvard University Press.

Lewis, David K. 1988. What experience teaches. Proceedings of the Russellian Society, Sydney, Australia.

Liu, Taosheng, Sean T. Stevens, and Marisa Carrasco. 2007. Comparing the time course and efficacy of spatial and feature-based attention. *Vision Research* 47 (1): 108–113.

Loftus, Elizabeth F. 2005. Planting misinformation in the human mind: A 30-year investigation of the malleability of memory. *Learning & Memory* 12 (4): 361–366.

Lombardo, Michael V., Bhismadev Chakrabarti, Edward T. Bullmore, Susan A. Sadek, Greg Pasco, Sally J. Wheelwright, and John Suckling, and the Mrc Aims Consortium, and Simon Baron-Cohen. 2010. Atypical neural self-representation in autism. *Brain* 133 (2): 611–624.

Lovibond, Peter F., Jean C. J. Liu, Gabrielle Weidemann, and Christopher J. Mitchell. 2011. Awareness is necessary for differential trace and delay eyeblink conditioning in humans. *Biological Psychology* 87 (3): 393–400.

Lu, Shena, and Shihui Han. 2009. Attentional capture is contingent on the interaction between task demand and stimulus salience. *Attention, Perception & Psychophysics* 71 (5): 1015–1026.

Luck, Steven J., Leonardo Chelazzi, Steven A. Hillyard, and Robert Desimone. 1997. Neural mechanisms of spatial selective attention in areas V1, V2, and V4 of macaque visual cortex. *Journal of Neurophysiology* 77 (1): 24–42.

Luck, Steven J., and Edward K. Vogel. 1997. The capacity of visual working memory for features and conjunctions. *Nature* 390 (6657): 279–281.

Lycan, William G. 1996. *Consciousness and Experience*. Cambridge, MA: MIT Press.

Mack, Arien, and Irvin Rock. 1998. *Inattentional Blindness*. Cambridge, MA: MIT Press.

Macphail, Euan M. 1998. *The Evolution of Consciousness*. Oxford: Oxford University Press.

Mangun, George R. 1995. Neural mechanisms of visual selective attention. *Psychophysiology* 32 (1): 4–18.

Marchetti, Giorgio. 2012. Against the view that consciousness and attention are fully dissociable. *Frontiers in Psychology* 3:36.

Marcus, Gary F. 2008. *Kluge: The Haphazard Construction of the Human Mind*. Boston, MA: Houghton Mifflin.

Markowitsch, Hans J., and Angelica Staniloiu. 2011. Memory, autonoetic consciousness, and the self. *Consciousness and Cognition* 20 (1): 16–39.

Marois, René, and Jason Ivanoff. 2005. Capacity limits of information processing in the brain. *Trends in Cognitive Sciences* 9 (6): 296–305.

Marois, René, Do-Joon Yi, and Marvin M. Chun. 2004. The neural fate of consciously perceived and missed events in the aftentional blink. *Neuron* 41 (3): 465–472.

Marr, David. 1980. Visual information processing: The structure and creation of visual representations. *Philosophical Transactions of the Royal Society of London. Series B, Biological Sciences* 290 (1038): 199–218.

Mason, Malia F., Michael I. Norton, John D. Van Horn, Daniel M. Wegner, Scott T. Grafton, and C. Neil Macrae. 2007. Wandering minds: The default network and stimulus-independent thought. *Science* 315 (5810): 393–395.

Mathôt, Sebastiaan, Clayton Hickey, and Jan Theeuwes. 2010. From reorienting of attention to biased competition: Evidence from hemifield effects. *Attention, Perception & Psychophysics* 72 (3): 651–657.

Mathôt, Sebastiaan, and Jan Theeuwes. 2011. Visual attention and stability. *Philosophical Transactions of the Royal Society of London. Series B, Biological Sciences* 366 (1564): 516–527.

Matzel, Louis D., and Stefan Kolata. 2010. Selective attention, working memory, and animal intelligence. *Neuroscience and Biobehavioral Reviews* 34 (1): 23–30.

Maunsell, John H., and Stefan Treue. 2006. Feature-based attention in visual cortex. *Trends in Neurosciences* 29 (6): 317–322.

McDowell, John Henry. 1994. *Mind and World*. Cambridge, MA: Harvard University Press.

Meeter, Martijn, Stefan Van der Stigchel, and Jan Theeuwes. 2010. A competitive integration model of exogenous and endogenous eye movements. *Biological Cybernetics* 102 (4): 271–291.

Melcher, David. 2011. Visual stability. *Philosophical Transactions of the Royal Society of London. Series B, Biological Sciences* 366 (1564): 468–475.

Merker, Bjorn. 2005. The liabilities of mobility: A selection pressure for the transition to consciousness in animal evolution. *Consciousness and Cognition* 14 (1): 89–114.

Merker, Bjorn. 2007a. Consciousness without a cerebral cortex: A challenge for neuroscience and medicine. *Behavioral and Brain Sciences* 30 (1): 63–81.

Merker, Bjorn. 2007b. Grounding consciousness: The mesodiencephalon as thalamocortical base. *Behavioral and Brain Sciences* 30 (1): 110–134.

Metzinger, Thomas. 2003. *Being No One: The Self-Model Theory of Subjectivity*. Cambridge, MA: MIT Press.

Meuwese, Julia D. I., Ruben A. G. Post, H. Steven Scholte, and Victor A. F. Lamme. 2013. Does perceptual learning require consciousness or attention? *Journal of Cognitive Neuroscience* 25 (10): 1579–1596.

Miller, George A. 1953. What is information measurement? *American Psychologist* 8 (1): 3–11.

Miller, George A. 1956. The magical number seven plus or minus two: Some limits on our capacity for processing information. *Psychological Review* 63 (2): 81–97.

Millikan, Ruth Garrett. 1990. The myth of the essential indexical. *Noûs* 24 (5): 723–734.

Millikan, Ruth Garrett. 2001. The myth of mental indexicals. In *Self-Reference and Self-Awareness*, ed. Andrew Brook and Richard C. DeVidi, 163–177. Amsterdam: J. Benjamins Pub.

Millikan, Ruth Garrett. 2004. *Varieties of Meaning*. Cambridge, MA: MIT Press.

Millikan, Ruth Garrett. 2005. *Language: A Biological Model*. New York: Oxford University Press.

Milner, A. David, and Melvyn A. Goodale. 1993. Visual pathways to perception and action. In *Progress in Brain Research*, ed. T. Philip Hicks, Stéphane Molotchnikoff and Taketoshi Ono, 317–337. Amsterdam: Elsevier.

Milner, A. David, and Melvyn A. Goodale. 1995. *The Visual Brain in Action*. 1st ed. Oxford: Oxford University Press.

Milner, A. David, and Melvyn A. Goodale. 2008. Two visual systems re-viewed. *Neuropsychologia* 46 (3): 774–785.

Mishkin, Mortimer, Leslie G. Ungerleider, and Kathleen A. Macko. 1983. Object vision and spatial vision: Two cortical pathways. *Trends in Neurosciences* 6:414–417.

Mitchell, Karen J., and Marcia K. Johnson. 2009. Source monitoring 15 years later: What have we learned from fMRI about the neural mechanisms of source memory? *Psychological Bulletin* 135 (4): 638–677.

Mitroff, Stephen R., Brian J. Scholl, and Karen Wynn. 2005. The relationship between object files and conscious perception. *Cognition* 96 (1): 67–92.

Mole, Christopher. 2008. Attention and consciousness. *Journal of Consciousness Studies* 15 (4): 86–104.

Mole, Christopher. 2011. *Attention is Cognitive Unison: An Essay in Philosophical Psychology*. New York: Oxford University Press.

Mole, Christopher, Declan Smithies, and Wayne Wu. 2011. *Attention: Philosophical and Psychological Essays*. Oxford: Oxford University Press.

Montemayor, Carlos. 2013. *Minding Time: A Philosophical and Theoretical Approach to the Psychology of Time*. Boston: Brill.

Montemayor, Carlos. 2014. Success, minimal agency, and epistemic virtue. In *Virtue Epistemology Naturalized: Bridges between Virtue Epistemology and Philosophy of Science*, ed. Abrol Fairweather. New York: Springer.

Montemayor, Carlos. (in press). Tradeoffs between the accuracy and integrity of autobiographical narrative in memory reconsolidation. *Behavioral and Brain Sciences*.

Montemayor, Carlos, Allison K. Allen, and Ezequiel Morsella. 2013. The seeming stability of the unconscious homunculus. *Sistemi Intelligenti* 3:581–600.

Moore, Cathleen M., and Howard E. Egeth. 1998. How does feature-based attention affect visual processing? *Journal of Experimental Psychology: Human Perception and Performance* 24 (4): 1296–1310.

Moore, Cathleen M., Steven Yantis, and Barry Vaughan. 1998. Object-based visual selection: Evidence from perceptual completion. *Psychological Science* 9 (2): 104–110.

Moran, Richard. 2001. *Authority and Estrangement: An Essay on Self-knowledge*. Princeton, NJ: Princeton University Press.

Most, Steven B. 2010. What's "inattentional" about inattentional blindness? *Consciousness and Cognition* 19 (4): 1102–1104.

Most, Steven B., Brian J. Scholl, Erin R. Clifford, and Daniel J. Simons. 2005. What you see is what you set: sustained inattentional blindness and the capture of awareness. *Psychological Review* 112 (1): 217–242.

Most, Steven B., Daniel J. Simons, Brian J. Scholl, Rachel Jimenez, Erin Clifford, and Christopher F. Chabris. 2001. How not to be seen: The contribution of similarity

References

and selective ignoring to sustained inattentional blindness. *Psychologica*. (1): 9–17.

Mudrik, Liad, Nathan Faivre, and Christof Koch. 2014. Information integration without awareness. *Trends in Cognitive Sciences* 18 (9): 488–496.

Mulckhuyse, Manon, and Jan Theeuwes. 2010. Unconscious attentional orienting to exogenous cues: A review of the literature. *Acta Psychologica* 134 (3): 299–309.

Nagel, Thomas. 1974. What is it like to be a bat? *Philosophical Review* 83 (4): 435–450.

Neisser, Ulric. 1967. *Cognitive Psychology*. New York: Appleton-Century-Crofts.

Neuman, Yair, and Ophir Nave. 2010. Why the brain needs language in order to be self-conscious. *New Ideas in Psychology* 28 (1): 37–48.

New, Joshua, Leda Cosmides, and John Tooby. 2007. Category-specific attention for animals reflects ancestral priorities, not expertise. *Proceedings of the National Academy of Sciences of the United States of America* 104 (42): 16598–16603.

Newell, Ben R., and David R. Shanks. 2014. Unconscious influences on decision making: A critical review. *Behavioral and Brain Sciences* 37 (01): 1–19.

Newsome, William T., and Edmond B. Paré. 1988. A selective impairment of motion perception following lesions of the middle temporal visual area (MT). *Journal of Neuroscience* 8 (6): 2201–2211.

Nichols, Shaun, and Todd Grantham. 2000. Adaptive complexity and phenomenal consciousness. *Philosophy of Science* 67 (4): 648–670.

Nietzsche, Friedrich W. [1887] 1974. *The Gay Science*. 1st ed. Translated by Walter A. Kaufmann. New York: Vintage Books.

Nisbett, Richard E., and Timothy D. Wilson. 1977. Telling more than we can know: Verbal reports on mental processes. *Psychological Review* 84 (3): 231–259.

Nissen, Mary J. 1985. Accessing features and objects: Is location special? In *Attention and Performance XI*, ed. Michael I. Posner and Oscar S. M. Marin, 205–219. Hillsdale, NJ: Lawrence Erlbaum Associates.

Noles, Nicholaus S., Brian J. Scholl, and Stephen R. Mitroff. 2005. The persistence of object file representations. *Perception & Psychophysics* 67 (2): 324–334.

Norman, Liam J., Charles A. Heywood, and Robert W. Kentridge. 2013. Object-based attention without awareness. *Psychological Science* 24 (6): 836–843.

Nunn, Chris. 2009. Editor's introduction: Defining consciousness. *Journal of Consciousness Studies* 16 (5): 5–8.

O'Craven, Kathleen M., Paul E. Downing, and Nancy Kanwisher. 1999. fMRI evidence for objects as the units of attentional selection. *Nature* 401 (6753): 584–587.

Oberauer, Klaus, and Laura Hein. 2012. Attention to information in working memory. *Current Directions in Psychological Science* 21 (3): 164–169.

Olivers, Christian N. L. 2008. Interactions between visual working memory and visual attention. *Frontiers in Bioscience* 13:1182–1191.

Palmer, Terry D., and Ashley K. Ramsey. 2012. The function of consciousness in multisensory integration. *Cognition* 125 (3): 353–364.

Palmiter, Richard D. 2011. Dopamine signaling as a neural correlate of consciousness. *Neuroscience* 198:213–220.

Parasuraman, Raja. 1998. *The Attentive Brain*. Cambridge, MA: MIT Press.

Park, Hyeong-Dong, and Catherine Tallon-Baudry. 2014. The neural subjective frame: From bodily signals to perceptual consciousness. *Philosophical Transactions of the Royal Society of London. Series B, Biological Sciences* 369 (1641): 20130208.

Parrott, Andrew C. 2013. MDMA, serotonergic neurotoxicity, and the diverse functional deficits of recreational 'Ecstasy' users. *Neuroscience and Biobehavioral Reviews* 37 (8): 1466–1484.

Pascual-Leone, Juan. 2006. Mental attention, not language, may explain evolutionary growth of human intelligence and brain size. *Behavioral and Brain Sciences* 29:19–21.

Pashler, Harold E. 1998. Visual search. In *Attention*, ed. Harold E. Pashler, 13–74. Hove: Psychology Press.

Payne, Jessica D., Matthew A. Tucker, Jeffrey M. Ellenbogen, Erin J. Wamsley, Matthew P. Walker, Daniel L. Schacter, and Robert Stickgold. 2012. Memory for semantically related and unrelated declarative information: The benefit of sleep, the cost of wake. *PLoS ONE* 7 (3): e33079.

Peacocke, Christopher. 2001. Does perception have a nonconceptual content? *Journal of Philosophy* 98 (5): 239–264.

Peelen, Marius V., and Sabine Kastner. 2014. Attention in the real world: Toward understanding its neural basis. *Trends in Cognitive Sciences* 18 (5): 242–250.

Perry, John. 1997. Indexicals and demonstratives. In *Companion to the Philosophy of Language*, ed. Robert Hale and Crispin Wright. Oxford: Blackwell.

Petersen, Steven E., and Michael I. Posner. 2012. The attention system of the human brain: 20 years after. *Annual Review of Neuroscience* 35:73–89.

References

Peterson, Matthew S., Arthur F. Kramer, and David E. Irwin. 2004. Covert shifts of attention precede involuntary eye movements. *Perception & Psychophysics* 66 (3): 398–405.

Poincaré, Henri. [1913] 1963. Why space has three dimensions. In *Mathematics and Science: Last Essays*, ed. John W. Bolduc. New York: Dover Publications.

Polger, Thomas, and Owen Flanagan. 2002. Consciousness, adaptation and epiphenomenalism. In *Consciousness Evolving*, ed. James H. Fetzer, 21–41. Amsterdam: John Benjamins Pub.

Pollen, Daniel A. 2003. Explicit neural representations, recursive neural networks and conscious visual perception. *Cerebral Cortex* 13 (8): 807–814.

Pöppel, Ernst. 1988. *Mindworks: Time and Conscious Experience*. 1st ed. Boston: Harcourt Brace Jovanovich.

Posner, Michael I. 1980. Orienting of attention. *Quarterly Journal of Experimental Psychology* 32 (1): 3–25.

Posner, Michael I. 1992. Attention as a cognitive and neural system. *Current Directions in Psychological Science* 1 (1): 11–14.

Posner, Michael I. 1994. Attention: The mechanisms of consciousness. *Proceedings of the National Academy of Sciences of the United States of America* 91 (16): 7398–7403.

Posner, Michael I. 2012. Attentional networks and consciousness. *Frontiers in Psychology* 3:64.

Posner, Michael I., Yoram Cohen, and Robert D. Rafal. 1982. Neural systems control of spatial orienting. *Philosophical Transactions of the Royal Society of London. Series B, Biological Sciences* 298 (1089): 187–198.

Posner, Michael I., and Steven E. Petersen. 1990. The attention system of the human brain. *Annual Review of Neuroscience* 13:25–42.

Posner, Michael I., Charles R. Snyder, and Brian J. Davidson. 1980. Attention and the detection of signals. *Journal of Experimental Psychology* 109 (2): 160–174.

Prinz, Jesse J. 2012. *The Conscious Brain: How Attention Engenders Experience*. New York: Oxford University Press.

Prinz, Wolfgang. 2012. *Open Minds: The Social Making of Agency and Intentionality*. Cambridge, MA: MIT Press.

Proust, Joëlle. 2007. Metacognition and metarepresentation: Is a self-directed theory of mind a precondition for metacognition? *Synthese* 159 (2): 271–295.

Proust, Marcel. 1922. *Remembrance of Things Past, Part One, Swann's Way*. Translated by C. K. Scott Moncrieff. New York: Henry Holt and Company.

Pylyshyn, Zenon W. 1965. Information available from two consecutive exposures of visual displays. *Canadian Journal of Psychology* 19 (2): 133–144.

Pylyshyn, Zenon W. 1980. Computation and cognition: Issues in the foundations of cognitive science. *Behavioral and Brain Sciences* 3 (01): 111–132.

Pylyshyn, Zenon W. 1984. *Computation and Cognition: Toward a Foundation for Cognitive Science.* Cambridge, MA: MIT Press.

Pylyshyn, Zenon W. 1989. The role of location indexes in spatial perception: A sketch of the FINST spatial-index model. *Cognition* 32 (1): 65–97.

Pylyshyn, Zenon W. 1999. Is vision continuous with cognition? The case for cognitive impenetrability of visual perception. *Behavioral and Brain Sciences* 22 (3): 341–365, discussion 66–423.

Pylyshyn, Zenon W. 2001. Visual indexes, preconceptual objects, and situated vision. *Cognition* 80 (1–2): 127–158.

Pylyshyn, Zenon W. 2003a. Return of the mental image: Are there really pictures in the brain? *Trends in Cognitive Sciences* 7 (3): 113–118.

Pylyshyn, Zenon W. 2003b. *Seeing and Visualizing: It's Not What You Think.* Cambridge, MA: MIT Press.

Pylyshyn, Zenon W. 2004. Some puzzling findings in multiple object tracking: I. Tracking without keeping track of object identities. *Visual Cognition* 11 (7): 801–822.

Pylyshyn, Zenon W. 2006. Some puzzling findings in multiple object tracking (MOT): II. Inhibition of moving nontargets. *Visual Cognition* 14 (2): 175–198.

Pylyshyn, Zenon W. 2007. *Things and Places: How the Mind Connects with the World.* Cambridge, MA: MIT Press.

Pylyshyn, Zenon W., Harry H. Haladjian, Charles E. King, and James E. Reilly. 2008. Selective nontarget inhibition in Multiple Object Tracking. *Visual Cognition* 16 (8): 1011–1021.

Pylyshyn, Zenon W., and Ron W. Storm. 1988. Tracking multiple independent targets: Evidence for a parallel tracking mechanism. *Spatial Vision* 3 (3): 179–197.

Quiroga, Rodrigo Quian. 2012. *Borges and Memory: Encounters with the Human Brain.* Cambridge, MA: MIT Press.

Raffone, Antonino, and Martina Pantani. 2010. A global workspace model for phenomenal and access consciousness. *Consciousness and Cognition* 19 (2): 580–596.

Raffone, Antonino, Narayanan Srinivasan, and Cees van Leeuwen. 2014a. The interplay of attention and consciousness in visual search, attentional blink and working memory consolidation. *Philosophical Transactions of the Royal Society of London. Series B, Biological Sciences* 369 (1641): 20130215.

Raffone, Antonino, Narayanan Srinivasan, and Cees van Leeuwen. 2014b. Perceptual awareness and its neural basis: Bridging experimental and theoretical paradigms. *Philosophical Transactions of the Royal Society of London. Series B, Biological Sciences* 369 (1641): 20130203.

Raymond, Jane E., Kimron L. Shapiro, and Karen M. Arnell. 1992. Temporary suppression of visual processing in an RSVP task: An attentional blink? *Journal of Experimental Psychology: Human Perception and Performance* 18 (3): 849–860.

Reichenbach, Hans. 1958. *The Philosophy of Space and Time*. New York, NY: Dover Publications.

Rensink, Ronald A. 2000. Seeing, sensing, and scrutinizing. *Vision Research* 40 (10–12): 1469–1487.

Rensink, Ronald A. 2002. Change detection. *Annual Review of Psychology* 53:245–277.

Rensink, Ronald A. 2008. On the applications of change blindness. *Psychologia* 51 (2): 100–106.

Rensink, Ronald A. 2009. Attention: Change blindness and inattentional blindness. In *Encyclopedia of Consciousness*, ed. William P. Banks, 47–59. New York: Elsevier.

Rensink, Ronald A. In press. A function-centered taxonomy of visual attention. In *Phenomenal Qualities: Sense, Perception, and Consciousness*, ed. Paul Coates and Sam Coleman. Oxford: Oxford University Press.

Rensink, Ronald A., J. Kevin O'Regan, and James J. Clark. 1997. To see or not to see: The need for attention to perceive changes in scenes. *Psychological Science* 8 (5): 368–373.

Rice, Heather J., and David C. Rubin. 2009. I can see it both ways: First- and third-person visual perspectives at retrieval. *Consciousness and Cognition* 18 (4): 877–890.

Ridder, William H. III, and Alan Tomlinson. 1997. A comparison of saccadic and blink suppression in normal observers. *Vision Research* 37 (22): 3171–3179.

Rock, Irvin, Christopher M. Linnett, Paul Grant, and Arien Mack. 1992. Perception without attention: Results of a new method. *Cognitive Psychology* 24 (4): 502–534.

Roelfsema, Pieter R., and Roos Houtkamp. 2011. Incremental grouping of image elements in vision. *Attention, Perception & Psychophysics* 73 (8): 2542–2572.

Roelfsema, Pieter R., Victor A. F. Lamme, and Henk Spekreijse. 1998. Object-based attention in the primary visual cortex of the macaque monkey. *Nature* 395 (6700): 376–381.

Rosenbaum, David A. 2002. Motor control. In *Stevens' Handbook of Experimental Psychology: Sensation and Perception*, ed. Harold E. Pashler and Steven Yantis, 315–339. New York: Wiley.

Rosenthal, David M. 1997. A theory of consciousness. In *The Nature of Consciousness: Philosophical Debates*, ed. Ned Block, Owen J. Flanagan, and Güven Güzeldere, 729–753. Cambridge, MA: MIT Press.

Rosenthal, David M. 2002. How many kinds of consciousness? *Consciousness and Cognition* 11 (4): 653–665.

Rosenthal, David M. 2008. Consciousness and its function. *Neuropsychologia* 46 (3): 829–840.

Rumelhart, David E., Paul Smolensky, James L. McClelland, and Geoffrey E. Hinton. 1986. Schemata and sequential thought processes in PDP models. In *Parallel Distributed Processing: Explorations in the Microstructure of Cognition*, ed. David E. Rumelhart and James L. McClelland, 7–57. Cambridge, MA: MIT Press.

Russell, Bertrand. 1921. *The Analysis of Mind*. London: G. Allen & Unwin Ltd.

Ryle, Gilbert. 1949. *The Concept of Mind*. Chicago: University of Chicago Press.

Sass, Louis A., and Josef Parnas. 2003. Schizophrenia, consciousness, and the self. *Schizophrenia Bulletin* 29 (3): 427–444.

Scarry, Elaine. 2001. *Dreaming by the Book*. Princeton, NJ: Princeton University Press.

Schechtman, Marya. 1994. The truth about memory. *Philosophical Psychology* 7 (1): 3–18.

Schmidt, Ralph E., Philippe Gay, Delphine Courvoisier, Francoise Jermann, Grazia Ceschi, Melissa David, Kerstin Brinkmann, and Martial Van der Linden. 2009. Anatomy of the White Bear Suppression Inventory (WBSI): A review of previous findings and a new approach. *Journal of Personality Assessment* 91 (4): 323–330.

Scholl, Brian J. 2001. Objects and attention: The state of the art. *Cognition* 80 (1–2): 1–46.

Scholl, Brian J., Zenon W. Pylyshyn, and Jacob Feldman. 2001. What is a visual object? Evidence from target merging in multiple object tracking. *Cognition* 80 (1–2): 159–177.

Schwitzgebel, Eric. 2011. *Perplexities of Consciousness*. Cambridge, MA: MIT Press.

Sears, Christopher R., and Zenon W. Pylyshyn. 2000. Multiple object tracking and attentional processing. *Canadian Journal of Experimental Psychology* 54 (1): 1–14.

Seth, Anil K., and Bernard J. Baars. 2005. Neural Darwinism and consciousness. *Consciousness and Cognition* 14 (1): 140–168.

Seth, Anil K., Bernard J. Baars, and David B. Edelman. 2005. Criteria for consciousness in humans and other mammals. *Consciousness and Cognition* 14 (1): 119–139.

Seth, Anil K., Zoltán Dienes, Axel Cleeremans, Morten Overgaard, and Luis Pessoa. 2008. Measuring consciousness: Relating behavioural and neurophysiological approaches. *Trends in Cognitive Sciences* 12 (8): 314–321.

Shannon, Claude Elwood, and Warren Weaver. 1949. *The Mathematical Theory of Communication*. Urbana, IL: University of Illinois Press.

Shin, Yun Kyoung, Robert W. Proctor, and E. John Capaldi. 2010. A review of contemporary ideomotor theory. *Psychological Bulletin* 136 (6): 943–974.

Shomstein, Sarah. 2012. Object-based attention: Strategy versus automaticity. *Wiley Interdisciplinary Reviews: Cognitive Science* 3 (2): 163–169.

Shomstein, Sarah, and Marlene Behrmann. 2006. Cortical systems mediating visual attention to both objects and spatial locations. *Proceedings of the National Academy of Sciences of the United States of America* 103 (30): 11387–11392.

Siegel, Susanna. 2002. The role of perception in demonstrative reference. *Philosophers' Imprint* 2 (1): 1–21.

Siegel, Susanna. 2010. *The Contents of Visual Experience*. New York: Oxford University Press.

Simons, Daniel J. 2000. Attentional capture and inattentional blindness. *Trends in Cognitive Sciences* 4 (4): 147–155.

Simons, Daniel J., and Christopher F. Chabris. 1999. Gorillas in our midst: Sustained inattentional blindness for dynamic events. *Perception* 28 (9): 1059–1074.

Simons, Daniel J., and Ronald A. Rensink. 2005. Change blindness: Past, present, and future. *Trends in Cognitive Sciences* 9 (1): 16–20.

Singer, Wolf. 2013. Cortical dynamics revisited. *Trends in Cognitive Sciences* 17 (12): 616–626.

Sloman, Aaron. 1978. *The Computer Revolution in Philosophy: Philosophy, Science, and Models of Mind*. Hassocks. Sussex: The Harvester Press Limited.

Smallwood, Jonathan, Florence J. M. Ruby, and Tania Singer. 2013. Letting go of the present: Mind-wandering is associated with reduced delay discounting. *Consciousness and Cognition* 22 (1): 1–7.

Smith, John Maynard, and Eörs Szathmáry. 1995. *The Major Transitions in Evolution*. New York: W.H. Freeman Spektrum.

Smithies, Declan. 2011. What is the role of consciousness in demonstrative thought? *Journal of Philosophy* 108 (1): 5–34.

Sosa, Ernest. 2007. *A Virtue Epistemology*. New York: Oxford University Press.

Sosa, Ernest. 2009. *Reflective Knowledge*. New York: Oxford University Press.

Spelke, Elizabeth. 1990. Principles of object perception. *Cognitive Science* 14 (1): 29–56.

Spelke, Elizabeth. 1994. Initial knowledge: Six suggestions. *Cognition* 50 (1–3): 431–445.

Spencer, Herbert. [1855] 1987. *The Principles of Psychology*. 3rd ed. 2 vols. New York: D. Appleton & Company.

Spencer, Kevin M., Paul G. Nestor, Olga Valdman, Margaret A. Niznikiewicz, Martha E. Shenton, and Robert W. McCarley. 2011. Enhanced facilitation of spatial attention in schizophrenia. *Neuropsychology* 25 (1): 76–85.

Sperling, George. 1960. The information available in brief visual presentation. *Psychological Monographs* 74 (11): 1–29.

Sperling, George. 1963. A model for visual memory tasks. *Human Factors* 5:19–31.

Sperling, George. 1989. Three stages and two systems of visual processing. *Spatial Vision* 4 (2–3): 183–207.

Stalnaker, Robert. 1984. *Inquiry*. Cambridge, MA: MIT Press.

Stöttinger, Elisabeth, and Josef Perner. 2006. Dissociating size representation for action and for conscious judgment: Grasping visual illusions without apparent obstacles. *Consciousness and Cognition* 15 (2): 269–284.

Striedter, Georg F. 2005. *Principles of Brain Evolution*. Sunderland, MA: Sinauer Associates.

Striedter, Georg F. 2006. Precis of principles of brain evolution. *Behavioral and Brain Sciences* 29:1–36.

Sutton, John. 2012. Memory. In *The Stanford Encyclopedia of Philosophy* (Winter 2012 Edition), ed. Edward N. Zalta. http://plato.stanford.edu/entries/memory/.

Swets, John A., Wilson P. Tanner, Jr., and Theodore G. Birdsall. 1961. Decision processes in perception. *Psychological Review* 68:301–340.

Tallon-Baudry, Catherine. 2012. On the neural mechanisms subserving consciousness and attention. *Frontiers in Psychology* 2:397.

Talsma, Durk, Daniel Senkowski, Salvador Soto-Faraco, and Marty G. Woldorff. 2010. The multifaceted interplay between attention and multisensory integration. *Trends in Cognitive Sciences* 14 (9): 400–410.

Tanskanen, Topi, Jussi Saarinen, Lauri Parkkonen, and Riitta Hari. 2008. From local to global: Cortical dynamics of contour integration. *Journal of Vision* 8 (7):15, 1–2.

References

Taylor, John G. 1997. Neural networks for consciousness. *Neural Networks* 10 (7): 1207–1225.

Taylor, John G. 2003. Paying attention to consciousness. *Progress in Neurobiology* 71 (4): 305–335.

Taylor, John G. 2007. CODAM: A neural network model of consciousness. *Neural Networks* 20 (9): 983–992.

Taylor, John G., and Nickolaos Fragopanagos. 2007. Resolving some confusions over attention and consciousness. *Neural Networks* 20 (9): 993–1003.

Taylor, John Henry. 2013. Is attention necessary and sufficient for phenomenal consciousness? *Journal of Consciousness Studies* 20 (11–12): 173–194.

Theeuwes, Jan. 1992. Perceptual selectivity for color and form. *Perception & Psychophysics* 51 (6): 599–606.

Theeuwes, Jan. 1993. Visual selective attention: A theoretical analysis. *Acta Psychologica* 83 (2): 93–154.

Theeuwes, Jan. 1994a. Endogenous and exogenous control of visual selection. *Perception* 23 (4): 429–440.

Theeuwes, Jan. 1994b. Stimulus-driven capture and attentional set: Selective search for color and visual abrupt onsets. *Journal of Experimental Psychology: Human Perception and Performance* 20 (4): 799–806.

Theeuwes, Jan. 2004. Top-down search strategies cannot override attentional capture. *Psychonomic Bulletin & Review* 11 (1): 65–70.

Theeuwes, Jan. 2010. Top-down and bottom-up control of visual selection. *Acta Psychologica* 135 (2): 77–99.

Theeuwes, Jan. 2013. Feature-based attention: It is all bottom-up priming. *Philosophical Transactions of the Royal Society of London. Series B, Biological Sciences* 368 (1628): 20130055.

Theeuwes, Jan, and Richard Godljn. 2002. Irrelevant singletons capture attention: Evidence from inhibition of return. *Perception & Psychophysics* 64 (5): 764–770.

Theeuwes, Jan, Arthur F. Kramer, and Paul Atchley. 1999. Attentional effects on preattentive vision: Spatial precues affect the detection of simple features. *Journal of Experimental Psychology: Human Perception and Performance* 25 (2): 341–347.

Theeuwes, Jan, Sebastiaan Mathôt, and Jonathan Grainger. 2013. Exogenous object-centered attention. *Attention, Perception & Psychophysics* 75 (5): 812–818.

Theeuwes, Jan, Stefan Van der Stigchel, and Christian N. L. Olivers. 2006. Spatial working memory and inhibition of return. *Psychonomic Bulletin & Review* 13 (4): 608–613.

Thompson, Kirk G., Keri L. Biscoe, and Takashi R. Sato. 2005. Neuronal basis of covert spatial attention in the frontal eye field. *Journal of Neuroscience* 25 (41): 9479–9487.

Tipper, Steven P., Jon Driver, and Bruce Weaver. 1991. Object-centred inhibition of return of visual attention. *Quarterly Journal of Experimental Psychology* 43 (2): 289–298.

Tomasello, Michael. 1999. *The Cultural Origins of Human Cognition*. Cambridge, MA: Harvard University Press.

Tomer, Raju, Alexandru S. Denes, Kristin Tessmar-Raible, and Detlev Arendt. 2010. Profiling by image registration reveals common origin of annelid mushroom bodies and vertebrate pallium. *Cell* 142 (5): 800–809.

Tononi, Giulio. 2004. An information integration theory of consciousness. *BMC Neuroscience* 5:42.

Tononi, Giulio. 2008. Consciousness as integrated information: A provisional manifesto. *Biological Bulletin* 215 (3): 216–242.

Tononi, Giulio. 2012. Integrated information theory of consciousness: An updated account. *Archives Italiennes de Biologie* 150 (2–3): 56–90.

Tononi, Giulio, and Christof Koch. 2008. The neural correlates of consciousness: An update. *Annals of the New York Academy of Sciences* 1124:239–261.

Tooby, John, and Leda Cosmides. 1995. Mapping the evolved functional organization of the mind and brain. In *The Cognitive Neurosciences*, ed. Michael S. Gazzaniga and Emilio Bizzi, 1185–1197. Cambridge, MA: MIT Press.

Tootell, Roger B., John B. Reppas, Kenneth K. Kwong, Rafael Malach, Richard T. Born, Thomas J. Brady, Bruce R. Rosen, and John W. Belliveau. 1995. Functional analysis of human MT and related visual cortical areas using magnetic resonance imaging. *Journal of Neuroscience* 15 (4): 3215–3230.

Treisman, Anne. 1960. Contextual cues in selective listening. *Quarterly Journal of Experimental Psychology* 12 (4): 242–248.

Treisman, Anne. 1964. Monitoring and storage of irrelevant messages in selective attention. *Journal of Verbal Learning and Verbal Behavior* 3 (6): 449–459.

Treisman, Anne. 1969. Strategies and models of selective attention. *Psychological Review* 76 (3): 282–299.

Treisman, Anne. 1982. Perceptual grouping and attention in visual search for features and for objects. *Journal of Experimental Psychology: Human Perception and Performance* 8 (2): 194–214.

Treisman, Anne. 1988. Features and objects: The fourteenth Bartlett memorial lecture. *Quarterly Journal of Experimental Psychology* 40 (2): 201–237.

References

Treisman, Anne. 1993. The perception of features and objects. In *Attention: Selection, Awareness, and Control: A Tribute to Donald Broadbent*, ed. Alan D. Baddeley and Lawrence Weiskrantz, 5–35. Oxford: Clarendon Press.

Treisman, Anne. 1996. The binding problem. *Current Opinion in Neurobiology* 6 (2): 171–178.

Treisman, Anne. 1998. Feature binding, attention, and object perception. *Philosophical Transactions of the Royal Society of London. Series B, Biological Sciences* 353 (1373): 1295–1306.

Treisman, Anne. 2006. How the deployment of attention determines what we see. *Visual Cognition* 14 (4–8): 411–443.

Treisman, Anne, and Garry Gelade. 1980. A feature-integration theory of attention. *Cognitive Psychology* 12 (1): 97–136.

Treisman, Anne, and Stephen Gormican. 1988. Feature analysis in early vision: Evidence from search asymmetries. *Psychological Review* 95 (1): 15–48.

Treisman, Anne, and Weiwei Zhang. 2006. Location and binding in visual working memory. *Memory & Cognition* 34 (8): 1704–1719.

Tremoulet, Patrice D., Alan M. Leslie, and D. Geoffrey Hall. 2000. Infant individuation and identification of objects. *Cognitive Development* 15 (4): 499–522.

Trick, Lana M., and Zenon W. Pylyshyn. 1993. What enumeration studies can show us about spatial attention: Evidence for limited capacity preattentive processing. *Journal of Experimental Psychology: Human Perception and Performance* 19 (2): 331–351.

Trick, Lana M., and Zenon W. Pylyshyn. 1994. Why are small and large numbers enumerated differently? A limited-capacity preattentive stage in vision. *Psychological Review* 101 (1): 80–102.

Tsotsos, John K., Scan M. Culhane, Yan Kei Wai Winky, Yuzhong Lai, Neal Davis, and Fernando Nuflo. 1995. Modeling visual attention via selective tuning. *Artificial Intelligence* 78 (1–2): 507–545.

Tsuchiya, Naotsugu, and Jeroen J. A. van Boxtel. 2013. Introduction to research topic: Attention and consciousness in different senses. *Frontiers in Psychology* 4:249.

Tsukiura, Takashi, and Roberto Cabeza. 2011. Shared brain activity for aesthetic and moral judgments: Implications for the Beauty-is-Good stereotype. *Social Cognitive and Affective Neuroscience* 6 (1): 138–148.

Tulving, Endel. 2002. Episodic memory: From mind to brain. *Annual Review of Psychology* 53:1–25.

Tversky, Amos. 1977. Features of similarity. *Psychological Review* 84 (4): 327–352.

Tye, Michael. 1995. *Ten Problems of Consciousness: A Representational Theory of the Phenomenal Mind.* Cambridge, MA: MIT Press.

Tye, Michael. 2003. *Consciousness and Persons: Unity and Identity.* Cambridge, MA: MIT Press.

Udell, Monique A. R., Nicole R. Dorey, and Clive D. L. Wynne. 2011. Can your dog read your mind? Understanding the causes of canine perspective taking. *Learning & Behavior* 39 (4): 289–302.

Ullman, Shimon. 1984. Visual routines. *Cognition* 18 (1–3): 97–159.

Ungerleider, Leslie G., and James V. Haxby. 1994. 'What' and 'where' in the human brain. *Current Opinion in Neurobiology* 4 (2): 157–165.

Ushitani, Tomokazu, Tomoko Imura, and Masaki Tomonaga. 2010. Object-based attention in chimpanzees (*Pan troglodytes*). *Vision Research* 50 (6): 577–584.

van Boxtel, Jeroen J. A., Naotsugu Tsuchiya, and Christof Koch. 2010. Consciousness and attention: On sufficiency and necessity. *Frontiers in Psychology* 1 (217).

Van der Burg, Erik, Christian N. L. Olivers, Adelbert W. Bronkhorst, and Jan Theeuwes. 2008. Pip and pop: Nonspatial auditory signals improve spatial visual search. *Journal of Experimental Psychology: Human Perception and Performance* 34 (5): 1053–1065.

Vandenbroucke, Annelinde R. E., Ilja G. Sligte, Adam B. Barrett, Anil K. Seth, Johannes J. Fahrenfort, and Victor A. F. Lamme. 2014. Accurate metacognition for visual sensory memory representations. *Psychological Science* 25 (4): 861–873.

Varela, Francisco J. 1996. Neurophenomenology: A methodological remedy for the hard problem. *Journal of Consciousness Studies* 3 (4): 330–349.

Vecera, Shaun P., and Martha J. Farah. 1994. Does visual attention select objects or locations? *Journal of Experimental Psychology. General* 123 (2): 146–160.

Verghese, Preeti. 2001. Visual search and attention: A signal detection theory approach. *Neuron* 31 (4): 523–535.

Vetter, Petra, and Albert Newen. 2014. Varieties of cognitive penetration in visual perception. *Consciousness and Cognition* 27 (0): 62–75.

Voss, Ursula, Karin Schermelleh-Engel, Jennifer Windt, Clemens Frenzel, and Allan Hobson. 2013. Measuring consciousness in dreams: The lucidity and consciousness in dreams scale. *Consciousness and Cognition* 22 (1): 8–21.

Wamsley, Erin J., and Robert Stickgold. 2010. Dreaming and offline memory processing. *Current Biology* 20 (23): R1010–R1013.

Wang, Ranxiao Frances, and Elizabeth Spelke. 2002. Human spatial representation: Insights from animals. *Trends in Cognitive Sciences* 6 (9): 376–382.

References

Wannig, Aurel, Liviu Stanisor, and Pieter R. Roelfsema. 2011. Automatic spread of attentional response modulation along Gestalt criteria in primary visual cortex. *Nature Neuroscience* 14 (10): 1243–1244.

Ward, Robert. 2013. Attention, evolutionary perspectives. In *Encyclopedia of the Mind*, ed. Harold E. Pashler, 53–56. Thousand Oaks, CA: Sage Publications.

Watson, Stephen E., and Arthur F. Kramer. 1999. Object-based visual selective attention and perceptual organization. *Perception & Psychophysics* 61 (1): 31–49.

Watson, Tamara L., and Bart Krekelberg. 2009. The relationship between saccadic suppression and perceptual stability. *Current Biology* 19 (12): 1040–1043.

Wegner, Daniel M. 1987. Transactive memory: A contemporary analysis of the group mind. In *Theories of Group Behavior*, ed. Brian Mullen and George R. Goethals, 185–208. New York: Springer-Verlag.

Wegner, Daniel M. 1994. Ironic processes of mental control. *Psychological Review* 101 (1): 34–52.

Wegner, Daniel M. 2003. The mind's best trick: How we experience conscious will. *Trends in Cognitive Sciences* 7 (2): 65–69.

Weiskrantz, Lawrence. 1996. Blindsight revisited. *Current Opinion in Neurobiology* 6 (2): 215–220.

Weiskrantz, Lawrence. 2009. *Blindsight: A Case Study Spanning 35 Years and New Developments*. 2nd ed. Oxford: Oxford University Press.

Westwood, David A., and Melvyn A. Goodale. 2011. Converging evidence for diverging pathways: Neuropsychology and psychophysics tell the same story. *Vision Research* 51 (8): 804–811.

Wheeler, Mary E., and Anne M. Treisman. 2002. Binding in short-term visual memory. *Journal of Experimental Psychology. General* 131 (1): 48–64.

Wiederman, Steven D., and David C. O'Carroll. 2013. Selective attention in an insect visual neuron. *Current Biology* 23 (2): 156–161.

Wiederman, Steven D., Patrick A. Shoemaker, and David C. O'Carroll. 2008. A model for the detection of moving targets in visual clutter inspired by insect physiology. *PLoS ONE* 3 (7): e2784.

Williams, Melonie, Pierre Pouget, Leanne Boucher, and Geoffrey F. Woodman. 2013. Visual-spatial attention aids the maintenance of object representations in visual working memory. *Memory & Cognition* 41 (5): 698–715.

Willingham, Daniel B., Joanna Salidis, and John D. E. Gabrieli. 2002. Direct comparison of neural systems mediating conscious and unconscious skill learning. *Journal of Neurophysiology* 88 (3): 1451–1460.

Wittgenstein, Ludwig. [1969] 1972. *On Certainty*. Edited by Gertrude Elizabeth M. Anscombe and George Henrik von Wright. New York: Harper Torchbook.

Wolfe, Jeremy M. 1994. Guided Search 2.0 - A revised model of visual search. *Psychonomic Bulletin & Review* 1 (2): 202–238.

Wolfe, Jeremy M. 2012. The binding problem lives on: Comment on Di Lollo. *Trends in Cognitive Sciences* 16 (6): 307–308.

Wolfe, Jeremy M., and Sara C. Bennett. 1997. Preattentive object files: Shapeless bundles of basic features. *Vision Research* 37 (1): 25–43.

Wolfe, Jeremy M., and Kyle R. Cave. 1999. The psychophysical evidence for a binding problem in human vision. *Neuron* 24 (1): 11–17.

Wolfe, Jeremy M., Kyle R. Cave, and Susan L. Franzel. 1989. Guided search: An alternative to the feature integration model for visual search. *Journal of Experimental Psychology: Human Perception and Performance* 15 (3): 419–433.

Wolfe, Jeremy M., and Todd S. Horowitz. 2004. What attributes guide the deployment of visual attention and how do they do it? *Nature Reviews. Neuroscience* 5 (6): 495–501.

Wright, Richard D., and Lawrence M. Ward. 2008. *Orienting of Attention*. Oxford: Oxford University Press.

Wu, Wayne. 2011. What is conscious attention? *Philosophy and Phenomenological Research* 82 (1): 93–120.

Wu, Wayne. 2014. *Attention*. New York: Routledge.

Wyart, Valentin, and Catherine Tallon-Baudry. 2008. Neural dissociation between visual awareness and spatial attention. *Journal of Neuroscience* 28 (10): 2667–2679.

Xu, Fei, and Susan Carey. 1996. Infants' metaphysics: The case of numerical identity. *Cognitive Psychology* 30 (2): 111–153.

Xu, Yaoda, and Marvin M. Chun. 2006. Dissociable neural mechanisms supporting visual short-term memory for objects. *Nature* 440 (7080): 91–95.

Yantis, Steven. 1992. Multielement visual tracking: Attention and perceptual organization. *Cognitive Psychology* 24 (3): 295–340.

Yantis, Steven. 1993. Stimulus-driven attentional capture and attentional control settings. *Journal of Experimental Psychology: Human Perception and Performance* 19 (3): 676–681.

Yantis, Steven. 2000. Goal-directed and stimulus-driven determinants of attentional control. In *Control of Cognitive Processes. Attention and Performance XVIII*, ed. Stephen Monsell and Jon Driver, 73–103. Cambridge, MA: MIT Press.

References

Yantis, Steven, and Anne P. Hillstrom. 1994. Stimulus-driven attentional capture: Evidence from equiluminant visual objects. *Journal of Experimental Psychology: Human Perception and Performance* 20 (1): 95–107.

Yantis, Steven, and John Jonides. 1984. Abrupt visual onsets and selective attention: Evidence from visual search. *Journal of Experimental Psychology: Human Perception and Performance* 10 (5): 601–621.

Yantis, Steven, and John Jonides. 1996. Attentional capture by abrupt onsets: New perceptual objects or visual masking? *Journal of Experimental Psychology: Human Perception and Performance* 22 (6): 1505–1513.

Yantis, Steven, Jens Schwarzbach, John T. Serences, Robert L. Carlson, Michael A. Steinmetz, James J. Pekar, and Susan M. Courtney. 2002. Transient neural activity in human parietal cortex during spatial attention shifts. *Nature Neuroscience* 5 (10): 995–1002.

Yantis, Steven, and John T. Serences. 2003. Cortical mechanisms of space-based and object-based attentional control. *Current Opinion in Neurobiology* 13 (2): 187–193.

Yeshurun, Yaffa, and Marisa Carrasco. 1998. Attention improves or impairs visual performance by enhancing spatial resolution. *Nature* 396 (6706): 72–75.

Zaidel, Dahlia W., and Marcos Nadal. 2011. Brain intersections of aesthetics and morals: Perspectives from biology, neuroscience, and evolution. *Perspectives in Biology and Medicine* 54 (3): 367–380.

Zeki, Semir, and Andreas Bartels. 1999. Toward a theory of visual consciousness. *Consciousness and Cognition* 8 (2): 225–259.

Zentall, Thomas R. 2005. Selective and divided attention in animals. *Behavioural Processes* 69 (1): 1–15.

Zhaoping, Li. 2008. Attention capture by eye of origin singletons even without awareness—A hallmark of a bottom-up saliency map in the primary visual cortex. *Journal of Vision* 8 (5): 1–18.

Zhaoping, Li. 2012. Gaze capture by eye-of-origin singletons: Interdependence with awareness. *Journal of Vision* 12 (2):17, 1–22.

Zmigrod, Sharon, and Bernhard Hommel. 2011. The relationship between feature binding and consciousness: Evidence from asynchronous multi-modal stimuli. *Consciousness and Cognition* 20 (3): 586–593.

Zmigrod, Sharon, Michiel Spapé, and Bernhard Hommel. 2009. Intermodal event files: Integrating features across vision, audition, taction, and action. *Psychological Research* 73 (5): 674–684.

Index

access conscious attention, 91–92, 105
  and blindsight, 114
  and cognitive integration, 113
  phenomenal and non-phenomenal, 108, 112, 119, 133
  and reality monitoring, 154, 173
  and social cognition, 135–137
access conscious normativity, 121
access monitoring, 132
  and epistemic normativity, 133
  and self-awareness, 134, 168
  action selection, 123, 125–127
  and the activity theorists, 138
  and conditioned training, 137
  and motor control, 129
activity theory, 135
aesthetic experiences, 163–164
ambiguous images, 8–10, 75
  and conscious attention, 143, 194
  duck-rabbit drawing, Necker cube, 9
attention
  all-or-none theory of, 45
  attenuating theory of, 45
  biased competition theory of, 47
  bottom-up, 43–49, 58, 59–60, 79, 188, 196–197, 205
  and consciousness, 33, 64–72, 78–80, 83
  covert, 34–35, 75, 189
  evolution of, 54–56, 81, 179–184, 185–186, 188, 196–197
  feature-based, 28–32, 56–60, 180–181, 184–185, 188–189, 200
  filter theory of, 44, 45
  flexible resource theory of, 41
  and iconic memory, 45, 90–91
  and inhibition, 42, 49, 53, 58, 62
  to intention, 167–168
  involuntary, 52, 118–119, 196, 197
  and memory, 30, 35, 39–41, 48, 50–53, 113, 118, 180–181, 189–190, 195, 198, 205
  neural structures of, 54–64, 185–186, 188–189
  object-based, 36–43, 46–47, 60–61, 69–70, 184, 190–191, 195, 205
  preattentive processes, 30, 38, 46, 47, 50, 147
  selective tuning model of, 50
  spatial, 32–36, 50, 59–61, 180, 184, 189
  spotlight theory of, 32, 34, 50, 59
  top-down, 43, 49–52, 58–60, 62, 75, 79, 188, 196–198
  voluntary, 52, 118–119, 166–167, 192, 195–197, 199
  zoom lens theory of, 32, 34
attentional blink, 69
attentional capture, 34, 47–48, 60, 62, 69
autobiographical memory, 20, 151–152, 156, 159, 166
  and narrative, 161, 171, 174

Baars, Bernard J., 13, 66, 74, 87, 89, 178, 183, 200, 202
Baddeley, Alan D., 51, 62, 87, 197
Baker, Lynn R., 20, 81–82
Bayne, Tim, 109–111, 114–115, 120, 134, 203
Beckett, Samuel, 164
Bernstein, Nicholai, A. 126, 137, 138
binding problem, 30, 38
binocular rivalry, 13, 80
blindsight, 16, 66, 67, 102, 113–114
blindsighter, super-duper, 103, 104, 122
  and epistemic agency, 126
Block, Ned, 7, 10–11, 86–88, 90–94, 103, 106, 114, 144, 200
Borges, Jorge L., 156
Broadbent, Donald, 30, 44–45
Brogaard, Berit, 67, 96, 117, 207
Bruya, Brian, 3, 52–54, 80, 198
Burge, Tyler, 101
Byrne, Alex, 92, 168, 221n11

Carrasco, Marisa, 10, 25, 32, 34, 75, 144
Campbell, John, 106–107, 152
Carruthers, Peter, 11, 94, 178, 187, 200, 202
central executive system, 51, 80, 126, 197
change blindness, 13, 67, 68f, 90
Chalmers, David J., 2, 7, 87, 93, 96, 109, 114, 131, 177, 183
Clark, Austen, 147–149
cocktail party effect, 44–45, 46, 64
cognitive integration, 99–100, 115, 124–125, 128–129, 148
  and epistemic access, 101–103
  and memory, 153, 159
  and Tononi's integrated information theory, 104, 193
cognitive impenetrability, 37, 43, 46, 128
Cohen, Michael A., 4, 65, 71, 117
collective memory, 159, 174

conceptual content, 8, 10, 144, 146, 194–195
congenital insensitivity to pain, 222n21
conjunctive attention, 111, 173
conscious attention, 3–4, 71–78, 141–176, 207–210, 213–215 (*see also* phenomenal conscious attention)
  and CAD, 5–6, 16–18, 91–92, 179
  evolution of, 178, 183–184, 198–205, 206f, 208–209
  and learning, 80–81, 145, 181, 183, 208–209
  object-based, 142–150
consciousness and attention dissociation (CAD), 4–6
  and its evolution (figure 5.1), 206
  spectrum of dissociation (figure 1.1), 6
consciousness
  access (definition), 7
  building block theory of, 203, 211, 212, 227n13
  creature, 19–20, 71–72, 203
  field theory of, 111, 173, 203
  first-order theories of, 21, 86, 92–95, 97–100
  higher-order theories of, 11, 86, 89, 92–100, 148
  and language, 81, 89, 111–112, 148, 187, 201
  and memory, 72, 80–81, 118–119, 151, 155–156, 203 (*see also* phenomenal traces)
  phenomenal (definition), 7
  and qualia, 7, 93
  state, 19
conscious field, 64, 111, 173
conscious voluntary attention, 157
continuity thesis of dreaming and daydreaming, 170–175
Cosmides, Leda, 3, 54, 56, 117, 178–180, 195
creativity, 150, 160, 163
crossmodal integration, 74, 143, 182–183

and attention, 192–194, 200–205, 210
and evolution, 181–182, 188, 193, 205
Csikszentmihalyi, Mihaly, 12, 138–139, 197

Dehaene, Stanislas, 65, 66, 74, 78–79, 87, 117–118, 178, 183, 202
demonstrative thoughts, 38, 104, 106–107, 111, 221n15, 227n8
Descartes, René, 1, 121, 160, 170–173
Dennett, Daniel C., 2, 3, 80, 82, 88–89, 99, 107, 178, 187, 200, 202
Di Lollo, Vincent, 66, 79, 118, 203
Diana, the huntress, 123–126
Dormashev, Yuri, 138
dreams, 113, 160–161, 163, 165, 168–176

early vision, 37–38, 45, 112, 190
effortless attention, 3, 52–53, 65, 77, 80, 136, 167, 176, 181, 197–198, 209, 213
effortless control, 138–139
empathic normativity, 136, 176
  and the enlightenment, 226n24
empathic understanding, 176
emotions
  and conscious attention, 149–150, 171
  and evolution, 180
  and memory, 150, 158–159
  and narrative, 161
  and phenomenally conscious enrichments, 163–164
  and self-awareness, 12, 142, 171
episodic memory, 151, 156
epistemic access, 102, 106, 108–109, 123, 125–126, 141, 148–150, 210
  and memory, 155–156, 159
epistemic agency, 85, 104–108, 119–126, 128, 153, 155
epistemic normativity, 85–86, 107–108, 112
  and access conscious normativity, 120–121, 124, 129, 131–132, 136

epistemic traces, 151–152, 160–161
  and density (Tolstoy's principle), 165
Evans, Gareth, 221n15
event files, 74–75, 192–193
evolution, 3, 16, 18
  of attention, 54–55, 81, 178, 179–184, 194
  of consciousness, 89, 183, 201, 204, 208–209
  of conscious attention, 21, 183–185, 199
expertise, 10, 52, 54, 126, 136, 143–145, 198

fear, 56, 136, 180, 214
feature integration, 29–31, 39, 148, 181, 191–192, 195
first-person point of view, 1, 19–20, 112
flow (experience of), 3, 12, 53, 80, 142, 167–168, 176
  and evolution, 196–198
  and expertise, 65, 136, 138–139, 213, 215
  and self-awareness, 136, 167–168, 213
Fodor, Jerry A., 146–147, 188, 194
Freud, Sigmund, 157, 174–175
functionalism, 21–22, 54, 88, 184
Funes's principle, 156

Gallistel, Charles R., 55, 74, 186, 193
Gestalt, 10, 11, 36, 38, 60, 128, 138, 145, 194
global broadcast, 74, 115, 148, 199, 202
global workspace, 73–74, 79–80, 87–88, 118, 183, 199, 202, 205
Goodale, Melvyn A., 30, 53, 59, 188
Gray, Jeffrey A., 87
Greco, John, 122–123, 128

Halbwachs, Maurice, 169, 174–175
hallucinations, 133, 160–161
Haugeland, John, 149

Hommel, Bernhard, 53, 69, 74, 138, 182, 192
Humphrey, Nicholas, 114, 199–200, 202, 214

icons, 145–149
identification of visual objects, 37, 39–41, 190
illusion, 133, 146
  accounts of consciousness as, 89, 187, 199–200, 202, 204
  and mental paint, 11
  of subjective perspective, 200
  visual, 65, 77, 128–129
imagery, 14, 160, 170, 173
imagination, 135, 150, 160, 176, 194
inattentional blindness, 68
individuation
  and icons, 147
  of visual objects, 37–38, 41–42, 190
information processing
  and bottlenecks, 41, 44
  and evolution, 183, 184–185
  theories of, 44–47
ironic processing, 157

Jackson, Frank, 131, 136, 163
James, William, 25, 27, 44, 130
Johnson, Marcia K., 132, 153, 169

Kahneman, Daniel, 37, 39, 41, 43, 138, 190
Kant, Immanuel, 11
Kentridge, Robert W., 15, 61, 67, 117, 207
Koch, Christof, 4, 12–13, 15, 18, 47, 58, 64, 66, 78–79, 104, 117, 178
Kornblith, Hilary, 123, 126
Kosslyn, Stephen M., 150
Krapp's Last Tape, 164

LaBerge, David A. F., 32, 49, 62, 66, 76, 82, 118

Lamme, Victor, 60, 62, 64, 72, 75, 197, 203
Lau, Hakwan C., 167
language of thought, 99, 148, 176
LeDoux, Joseph, 180
Leontiev, Aleksei N., 138
Lewis, David K., 131, 137
Lloyd, expert pianist and writer, 64–65, 138, 167, 197, 213
Loftus, Elizabeth F., 160
Lycan, William G., 11, 94

Maria, the super-duper blindsighter, 103–105, 107, 122
  and Diana, the huntress, 126
Mary, the neuroscientist, 131–132, 136, 163
McDowell, John H., 221n15, 227n9
mental paint, 7, 10–11, 144, 218n1
Millikan, Ruth G., 20, 137, 149
modularity, 28, 105, 188
motor control, 14, 53, 55, 56, 65, 100, 123–129, 133
multimodal integration (*see* crossmodal integration)
multiple object tracking (MOT), 38–39
mysterianism, 21–22

Nagel, Thomas, 2, 7, 200, 207
narrative, 150–152, 155, 157–161, 163–166, 169, 171, 174–175
neural structures
  and attention, 35, 43–44, 47, 54–64, 117–119, 185–189, 196
  and consciousness, 66, 72–73, 75, 77, 88, 117–119, 187, 196, 203, 207
  dorsal and ventral pathways, 30, 58–59, 188, 207
  illustration of, 57f
  visual cortex, 30, 55–61, 75, 79, 188, 191, 218n2
Nietzsche, Friedrich W., 80, 81

object files, 31, 39–41, 43, 74–75, 190–192
oneiric traces (dream memories), 169, 175
overflow argument, 90, 106, 108

pain, 72, 108, 115, 120, 130, 133–136, 149, 187, 205, 207
pain blob, 115–116
Peacocke, Christopher, 149
Perry, John, 38, 227n8
phenomenal conscious attention, 91–92, 106, 108, 110, 114–115, 133, 159
  in dreams, 173
phenomenal contrasts, 8, 26, 75, 144
phenomenal integration, 128, 150, 161, 164, 174
phenomenalism
  general, 92–93, 96
  about memory, 159–160
phenomenally conscious normativity, 129
phenomenal self-awareness, 13, 120–121, 134
phenomenal traces, 159–163, 165–166, 169–171, 174
Poincaré, Henri, 218n7
pop-out effect, 29, 47
postvoluntary attention, 138
Prinz, Jesse, J., 2, 3, 15, 71, 89, 93, 112–114, 118, 202
Prinz, Wolfgang, 119, 121, 135, 137
Proust, Joëlle, 128
Proust, Marcel, 162
Proustian flooding, 162–165
post-traumatic stress disorder (PTSD), 158, 164
Pylyshyn, Zenon W., 37–39, 42, 44, 111, 128, 144, 146, 190, 192

reality monitoring, 132–133, 153–154, 158–161, 163–164, 168–170, 172–175

recognitional phenomenal self-awareness, 134
recognitional self-awareness, 134, 137–138
reductionism, 21–22, 89, 106
Reichenbach, Hans, 224n6
Rensink, Ronald A., 14, 46, 67, 178, 184, 188
resemblance metrics, 146
Rosenthal, David M., 11, 94, 178
Rumelhart, David E., 154
Russell, Bertrand, 134, 152

Scarry, Elaine, 150
self
  phenomenal, 13, 18, 120–121, 130–131, 134
  recognitional, 13, 18, 120–121, 130
self-awareness
  and conscious attention, 76, 82, 136, 166, 175, 195–198
  and epistemic agency, 105, 119–120
  mirror test for, 12
  and normativity, 130, 134
self-knowledge, 130–131, 136, 168
selfless perspective, 161
semantic memory, 159, 181
semantic specificity (of images), 10, 143
Siegel, Susanna, 38, 96, 144
signal detection, 46, 186
Smithies, Declan, 107–108
Sosa, Ernest, 122–123, 127, 172–173
Spelke, Elizabeth, 38, 55, 82, 184
Spencer, Herbert, 177
Sperling, George, 44–45, 90
subjectivity, 19–20, 52, 76, 82, 98, 112, 134, 204, 207, 213
subsumption, 100, 109–111, 114–115, 120
  and memory, 161
  and oneiric attention, 173
  and self-knowledge, 135
subsumptive attention, 111

Tallon-Baudry, Catherine, 4, 15, 82, 118–119, 121, 131, 137, 226n3
temporal perspective, 158–159, 164, 166
Tolstoy's principle, 165
Tononi, Giulio, 14, 18, 66, 104, 183, 193
Tooby, John, 3, 54–56, 117, 178–180, 195, 196
transactive memory, 159
transparency method, 168, 175
Treisman, Anne, 30–31, 33, 37–41, 45–46, 49–50, 189–190, 192
Tsotsos, John, 50
Tsuchiya, Naotsugu, 4, 15, 65, 66, 71, 78–79, 117
Tulving, Endel, 156
Tye, Michael, 221n15, 222n19

Ullman, Shimon, 38, 47–48, 49
Urbach-Wiethe disease, 136

Varela, Francisco J., 21–22
value (moral and aesthetic), 132, 135, 150
visual indexing, 38–39, 41, 191
visualization, 144, 150
visual routines, 38, 48, 137
visual search, 28, 38, 49–50, 185
visual stability, 70–71

Ward, Robert, 54, 59, 117, 177–178, 180, 185–186
Wegner, Daniel M., 157, 159
Weiskrantz, Lawrence, 67
Wittgenstein, Ludwig, 152–153
Wu, Wayne, 3, 73, 121